Two Industries in Roman Lusitania

Mining and Garum Production

J. C. Edmondson

BAR International Series 362
1987

B.A.R.

5, Centremead, Osney Mead, Oxford OX2 0DQ, England.

BAR -S362, 1987: Two Industries in Roman Lusitania'

Price £20.00 post free throughout the world. Payments made in dollars must be calculated at the current rate of exchange and $8.00 added to cover exchange charges. Cheques should be made payable to B.A.R. and sent to the above address.

ISBN 0 86054 469 9

For details of all new B.A.R. publications in print please write to the above address. Information on new titles is sent regularly on request, with no obligation to purchase.

Volumes are distributed from the publisher. All B.A.R. prices are inclusive of postage by surface mail anywhere in the world.

Printed in Great Britain

TOWNS AND ROADS OF LUSITANIA

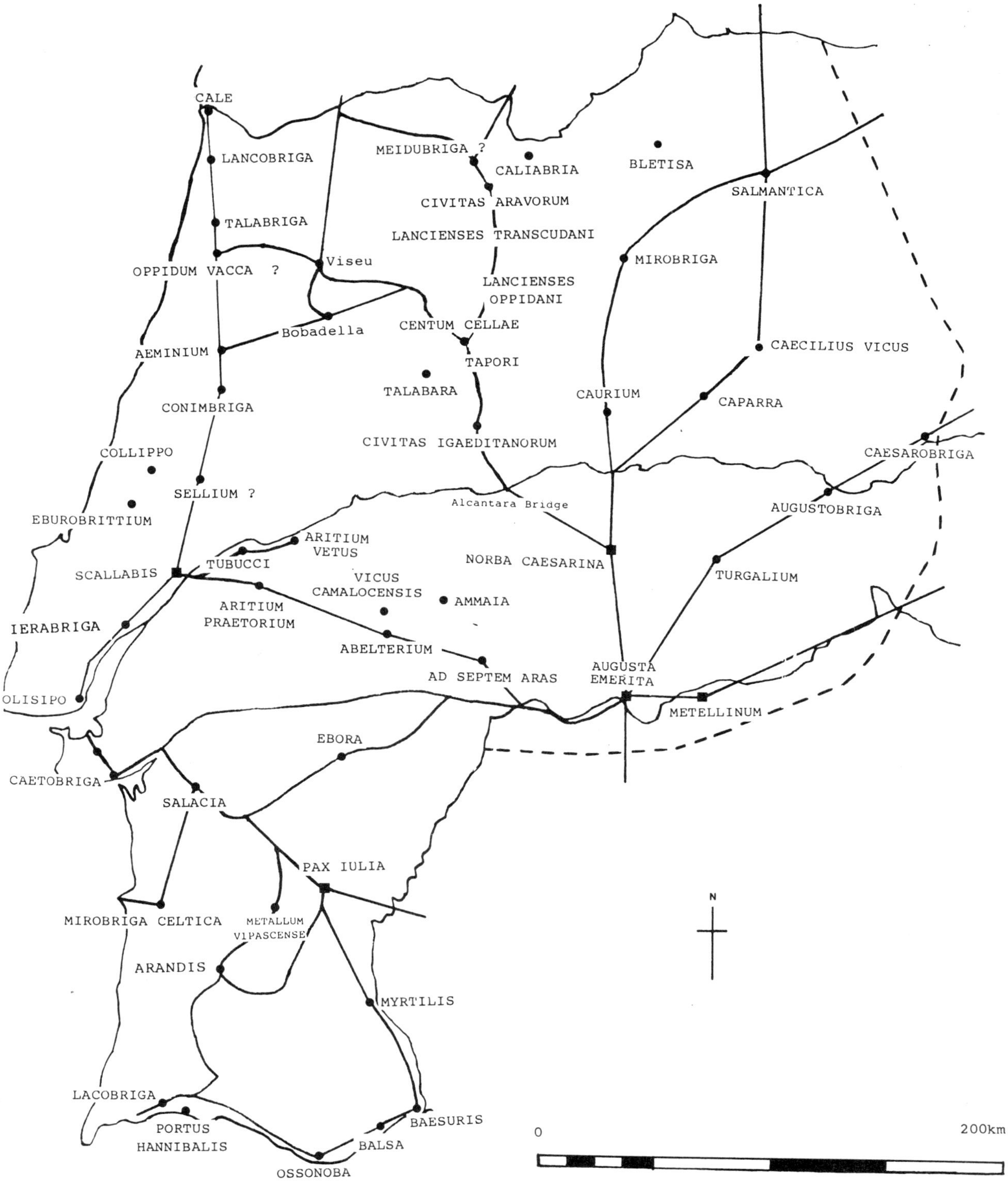

■ = colony ● = other town

TWO INDUSTRIES IN ROMAN LUSITANIA :

MINING AND GARUM PRODUCTION

TABLE OF CONTENTS

LIST OF FIGURES AND TABLES

PREFACE

The research on which this work is based was carried out in many libraries and museums not only in this country, but also in Spain, Portugal and France. I am, therefore, indebted to the staff of all those institutions. I am especially grateful to the University Library, to the Faculty of Classics and Museum of Classical Archaeology and to the Fitzwilliam Museum in Cambridge; to the Bodleian and the Ashmolean Museum in Oxford; to the British Library and the Joint Library of the Hellenic and Roman Societies in London; to Professor J.M. Blazquez, Drs. J. Arcé and F.J. Sanchez Palencia at the Instituto Rodrigo Caro and to Professor H. Schubart, Dr. W. Trillmich and all his colleagues at the Deutsches Archäologisches Institut in Madrid; to Professor J. de Alarcão not only for allowing me to work at the Instituto de Arqueologia at Coimbra, but for all his encouragement and help; to Professor Robert Étienne and his colleagues for all their generous help and hospitality during three periods spent working at the Centre Pierre Paris in Bordeaux.

Secondly, I am grateful to the following for all the financial assistance that was made available for the researching of this work : the Department of Education and Science, the Spanish Government for their kind grant of a Studentship, the British Academy and Christ's College, Cambridge.

Thirdly, I am indebted to many individuals who have helped me : to Sylvia Sylvester, who kindly typed early drafts of my work, to Esmond Rand and John Charnley, who helped to introduce me to the secrets of word processing at Dulwich, to various friends for their help with languages.

My greatest debt is to all those who have freely given of their advice and commented on various drafts of my work : to Professors Jorge de Alarcão, Robert Étienne and Claude Domergue, to Peter Garnsey, Tom Gallant, to Professor Michael Crawford, who has constantly fired my interest in Ancient History, advised me, read my work at various stages and tempered some of my rasher ideas, in addition to all his kind hospitality at home and abroad; to Professor John Crook, who was tireless and extremely generous in supervising the research on which this work is based; and finally, to my wife, who has been a great source of help and encouragement.

The abbreviations used throughout the work on those used in *L'Année Philologique*.

The following may be less familiar to readers :

AEA = *Archivo español de Arqueologia*
AP = *O Arqueologo Português*

Finally, this book is dedicated, as a small token of thanks for all their encouragement over many years, to my parents :

PARENTIBUS MEIS

CARISSIMIS INDULGENTISSIMISQUE.

CHAPTER ONE

AIMS, SOURCES, METHODS

Much recent work on the Roman economy has concentrated on the supply of vital resources, and especially foodstuffs, to the urban centres of the Roman Empire. Of these cities Rome has received the lion's share of attention, just as it once received the lion's share of the resources exported from the provinces of its Empire. More is now known about the areas of supply, the mechanisms and the personnel involved in the transfer and the special requirements of the large consumer centres. The need to keep a large standing army supplied with food and metals, especially iron, on the frontiers has also been recognized. Of the vital resources grain, wine and olive oil have received most attention. On olive oil, increased archaeological fieldwork in southern Spain has revealed much about the organization of production and has helped to elucidate the origin of many of the oil producers already attested on amphora stamps from the enormous dump of amphorae at Monte Testaccio in Rome. (1)

However, to date there have not been many attempts to set the production processes involved in their provincial context, to assess their importance to the local, as opposed to the Empire-wide, economy. Since land, and its exploitation, was clearly the predominant element in the economy of the Roman Empire, it will be important to establish the extent to which "industrial" processes were integrated with, or divorced from, agriculture. The production of wine and olive oil was by its very nature an integral part of the agricultural calendar. This study will concentrate on two products not so immediately and obviously linked to the exploitation of the land - metals and fish products.

The importance of the first needs little demonstration, since the demand for metals of all kinds has been great throughout history and has given an economic, and political, importance to those areas of the Mediterranean world blessed with mineral resources. As for the variety of fish sauces and conserves produced in the Roman period, although they do not appear at first sight to be what may be termed a staple, their ubiquity in Roman diet as condiments, as well as their use in medicine, and the widely differing grades of product (suggesting a wide potential market for them) make them a foodstuff worthy of investigation. How was production organized ? What role did it play in the provincial economy ? Can anything be deduced about the scale of production ? How was surplus produce disposed of ?

Since I want to study production processes on the local, as opposed to Empire-wide, level, I have decided to concentrate on a single Roman province, Lusitania, the most westerly of the three Iberian provinciae. Roman Spain, until recently, has not attracted much scholarly interest, at least in the English-speaking world; Roman Lusitania even less, perhaps because it covered areas of both Spain and Portugal, and native Iberian scholars tend to concentrate on the antiquities of their own countries. Lusitania, however, was rich in both raw materials under discussion, metals and fish, but in studies of the economy of Roman Spain its independent contribution is often overlooked. Thus studies on Roman mining tend to concentrate on the gold of the north-west or the silver of the Sierra Morena district of Baetica, while for fish products New Carthage and especially the Baetican coastal region receive far more attention. It is the aim of this study to investigate in detail the production processes involved in mining and fish-processing in just one province, rather than to present a composite picture drawing on material from all over the Iberian peninsula.

The chance of success of any investigation into the economy of a Roman province is, as ever, limited by the quality and quantity of the available data. Ancient literary evidence exists for both mining and fish-processing in Lusitania; it will be discussed, but unfortunately it is of disappointingly limited scope. Archaeological evidence is much more revealing, since in theory at least it should allow us to fix with greater precision the exact location of the resources exploited in Roman times and the periods in which a particular site was operating. Numismatic evidence is more difficult to handle securely. The discovery of an isolated coin at a given archaeological site is often immediately encouraging, but it does not necessarily follow that the site was in operation at the very time of the coin's minting, since coins often remained in circulation for long periods. Coin hoards are on the whole better, since it is legitimate to assume that they were buried soon after the date of the most recent coin(s). Inscriptions should tell us something about the personnel active in the production areas, but a large majority of the texts defy precise dating.

Finally, use has been made in this study of comparative evidence, in an attempt to fill some of the gaps left by the extremely partial nature of the ancient evidence. Ideally, records from the pre-industrial age are preferable, but occasionally material from the more recent past has been introduced to fill out the picture. Such comparative evidence can, of course, never be used to prove that a particular practice existed in the Roman period, but it can suggest possibilities which may then be tested against such ancient evidence as does survive. The techniques of Roman and, say, Spanish American mining were sufficiently analogous to make consideration of the latter worthwhile; similarly, the technology of fishing and fish-processing has only recently developed to a significant degree from the levels achieved in the Roman period; in some parts of the Third World the techniques even recently practised differ very little from those of Roman times.

There are also further limitations specific to the archaeological, epigraphic and numismatic material from Lusitania. First, archaeological work has not been carried out with the same intensity across the entire province, which means that there are certain gaping holes in the archaeological record. Thus distribution maps in particular may well chart not the real distribution of any chosen item in the Roman period, but merely the areas of the most intensive modern archaeological fieldwork. It also makes arguments from silence especially dangerous. The recent division of Portugal into three archaeological regions centred on Oporto (the North), Coimbra (the Centre) and Evora (the South) and the work of the archaeological units for the Spanish provinces of Badajoz, Cáceres and Toledo has done much to ensure a more even spread of interest. But problems still exist : for example, in the province of Salamanca archaeologists are still more interested in the pre- and proto-historic, rather than Roman, periods. (2)

Secondly, the publication of excavations, inscriptions and coins/coin hoards has not been as comprehensive and sophisticated as in some other parts of the Roman Empire. Much recent progress has been made : the publication of the Franco-Portuguese excavation of Conimbriga is exemplary and has become the model for more recent site reports. (3) Annual summaries of excavations are now published both in Spain and Portugal. (4) But many of the sites here under discussion were discovered in the nineteenth or early twentieth century, when archaeologists had more regard for eye-catching artefacts than for a careful stratigraphy or even a systematic inventory of finds. Thus many sites defy close dating. Further specific problems surround the identification of Roman mines, which are treated in detail below. (5)

As for inscriptions the second volume of the Corpus Inscriptionum Latinarum (CIL) and its supplement were published in 1869 and 1892 respectively and so are now hopelessly out of date. A revision programme is well underway and another supplement is promised within the next decade. Inscriptions discovered after 1892 have been published, if at all, in a great variety of local periodicals, some often badly misread. (6) The collection of inscriptions from the entire peninsula published by Vives in 1971, Inscripciones Latinas de la España romana (ILER), is unreliable. A catalogue of inscriptions from the province of Cáceres appeared in 1977, but the most significant recent advance in Lusitanian epigraphy has been provided by d'Encarnação with his splendidly thorough and well-presented catalogue of, and commentary on, all the inscriptions from the southern part of the province : Inscrições romanas do Conventus Pacensis (IRCP). (7) Finally, coin hoards have not received sufficiently critical attention until recently, since much of the material is in private hands and often remains unpublished. (8)

In studies of other provinces of the Roman Empire archaeological, epigraphic and numismatic material has been used to great advantage, not least because important methodological refinements have been developed to deal critically with this kind of evidence. Despite the views of some sceptics, non-literary evidence has much to reveal of the economic and social history of the ancient world. (9) The quantity and quality of such material now becoming available from the Iberian peninsula makes an overview highly desirable, so that its economic and social implications may be fully discussed. By a judicious blending of all types of evidence and by using methods and avenues of approach adopted in the study of other Roman provinces, it is hoped that new light can be thrown on the workings of two industries in Lusitania : namely, mining and garum production. The catalogues of sites and finds (for example, Appendices I, III, V and VI) and indeed the general conclusions drawn in this study must be regarded as provisional. It is hoped that further discoveries may help to confirm or modify the conclusions reached. The discussion may also serve to highlight areas where further archaeological work may prove beneficial.

NOTES : CHAPTER ONE

1. In general see M.I. Finley, The Ancient Economy, London, 1973; G.E. Rickman, The Corn Supply of Ancient Rome, Oxford, 1980; P. Garnsey, K. Hopkins & C.R. Whittaker (ed.), Trade in the Ancient Economy, London, 1983; K. Hopkins, 'Economic growth and towns in classical antiquity' in P. Abrams & E.A. Wrigley (ed.), Towns in Societies, Cambridge, 1978, 35-77; id., 'Taxes and trade in the Roman Empire', JRS, 70, 1980, 101-25; id., 'Models, ships and staples' in P. Garnsey & C.R. Whittaker (ed.), Trade and Famine in Classical Antiquity (= Cambridge Philological Society, Supplementary Volume 8), Cambridge, 1983, 84-109.
On supplying the Roman army : P. Middleton, 'Army supply in Roman Gaul' in B.C. Burnham & H.B. Johnson (ed.), Invasion and Response : the case of Roman Britain (B.A.R. British Series, 73), Oxford, 1979, 81-97; J. Remesal Rodriguez, La annona militaris y la exportación de aceite bético a Germania, Madrid, 1986.
On wine : A. Tchernia, Le vin de l'Italie romaine : essai d'histoire économique d'après les amphores (Bibl. des Écoles françaises d'Athènes et de Rome, 261), Rome, 1986.
On Baetican olive oil : T.R.S. Broughton, 'Oil producing estates in Roman Baetica', Revista de la Universidad Complutense, 18, 1979 (= Homenaje a Garcia y Bellido, IV), 153-68; J. Remesal Rodriguez, 'La economia oleicola bética : nuevas formas de analisis', AEA, 50-51, 1977-78, 87-142 & op.cit.; E. Rodriguez Almeida, Il Monte Testaccio : ambiente, storia, materiali, Rome, 1984; and note the proceedings of two recent congresses : J.M. Blazquez & J. Remesal (ed.), Producción y comercio del aceite en la antiguedad I (Madrid, 1980), Madrid, 1981; II (Sevilla, 1982), Madrid, 1983.

2. For a useful annual summary of archaeological work for Portugal see Informação arqueologica, published in Lisbon by the Ministry of Culture, Patrimonio Cultural since 1977-8; for Spain see Arqueologia. Memoria de las actuaciones programadas, published by the Ministry of Culture, Madrid, since 1979.

3. J. de Alarcão & R. Etienne (ed.), Fouilles de Conimbriga [= FC], 7 vols., Paris, 1974-1979.

4. See note 2.

5. See Chapter Three, pp. 27-28.

6. For inscriptions published after CIL see EE VIII & IX; Hispania Antiqua Epigraphica [= HAEp], an epigraphic supplement to Archivo Espanol de Arqueologia, published in Madrid from 1950 to 1966-69. For inscriptions from Portugal recent finds are often published in the journal Conimbriga, published by the Instituto de Arqueologia at the University of Coimbra, which since 1982 has an epigraphic supplement : Ficheiro Epigrafico.

7. J. Vives, Inscripciones latinas de la España romana, 2 vols., Barcelona, 1971 [= ILER]; R. Hurtado de San Antonio, Corpus provincial de inscripciones latinas (Cáceres), Cáceres, 1977 [= CPILC]; J. d'Encarnação, Inscrições romanas do Conventus Pacensis : subsidios para o estudo da romanização, Coimbra, 1984 [= IRCP].

8. For Lusitania see M. de Castro Hipolito, 'Dos tesouros de moedas romanas em Portugal', Conimbriga, 2-3, 1960-61, 1-166; I. Pereira, J.-P. Bost, J. Hiernard, FC III : Les Monnaies, Paris, 1974. In general see the journal Numisma.

9. See the comments of M.W. Frederiksen, 'Theory, evidence and the ancient economy', JRS, 65, 1975, 164-71.

CHAPTER TWO

THE PROVINCE OF LUSITANIA

The province of Lusitania was created by Augustus when he divided Hispania Ulterior into two new provinces, Baetica, which became a province nominally controlled by the Senate, and Lusitania, an imperial province. The colony of Augusta Emerita was founded on the site of the modern Mérida for the settlement of veterans from P. Carisius' army after the campaigns against the Asturians in 26-25 B.C. and became the capital of the new province of Lusitania. The date of this reorganization is much disputed, but does not concern us here. Later in his reign Augustus altered the frontiers of Lusitania, transferring some territory in the north to the province of Hispania Tarraconensis and removing all legions from Lusitania. Thus, although it was an imperial province, there was no Roman military garrison and so the governor was only of propraetorian rank. Tarraconensis became the only province with an army, based at León. There may also have been further small alterations in the extreme east under Vespasian. (1)

There are problems, however, over the exact limits of Lusitania, which need outlining at the outset. The northern border with Tarraconensis can securely be fixed as the river Durius (modern Duero/Douro). (2) The literary sources concur in isolating the river Anas (modern Guadiana) as the southern limit, and thus the border with Baetica. (3) However, several factors conspire to cast doubt on the accuracy of this assertion. (4)

FIGURE 2.1 MAP OF IBERIAN PENINSULA, showing TERRAIN, LIMITS OF LUSITANIA and CONVENTUS BOUNDARIES

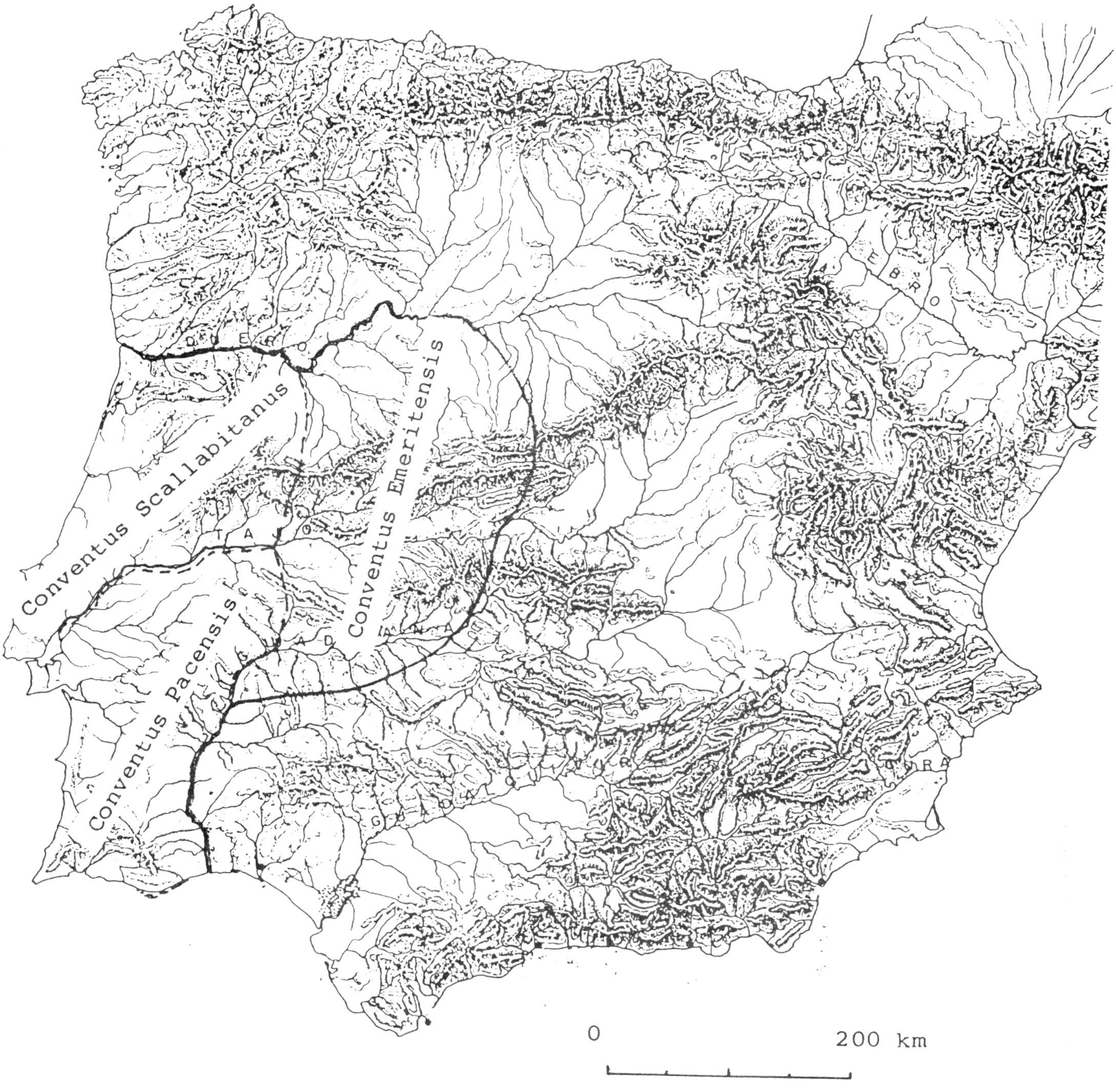

The first problem concerns the territory of Emerita. (5) Frontinus, a Roman consul and writer of agrimensorial treatises, seems to imply in the following passage that the territory extended on both sides of the river : (6)

> scio in Lusitania, finibus Emeritensium, non exiguam per mediam coloniae perticam ire flumen Anam, circa quod agri sunt adsignati.

This is confirmed by recent field survey, which has shown that land was centuriated for allotment to veterans not only to the north of the river, but also to the south (especially around Almendralejo in the fertile Tierra de Barros region). (7) The centuriated zone south of the river extends for thirty-five kilometres almost as far south as Los Santos de Maimona (to the south of Villafranca de los Barros). This would seem to place the provincial boundary in the region of the Sierra de Zafra, where lies the watershed between the Guadiana and the Guadalquivir. (8)

In addition, two termini Augustales have been discovered relating to the territory of Emerita. One was found at Montemolin (100 km. south of Emerita), the other at Valdecaballeros (120 km. to the east of Emerita). (9) The first marked the "Termin(u)s Aug(ustalis) pra(torum)/col(oniae) Aug(ustae) Emeritae", the second the "Augustalis Ter/minus c(olonorum) c(oloniae) C(laritatis) Iul(iae)/Ucubitanor(um)/inter Aug(ustanos) Emer(itenses)". The wording of the latter suggests that the colony of Ucubi (modern Espejo, to the south of the Guadalquivir) owned some land within the territory of Emerita. (10) The territory, therefore, extended at least this far east to the north of the Guadiana.

The interpretation of the text from Montemolin is less clear-cut. It could imply either that the territory stretched even further south than the centuriated land already mentioned; or that there was a group of colonists of Emerita who owned an enclave of land within the territory of a Baetican community - perhaps Curiga (modern Monasterio, 25 km. south-west of Montemolin). The land owned by the colonists of Emerita in this area is described as pasture land (prata) and so may have consisted merely of winter pasture for transhumant flocks. (11) Neither stone necessarily concerns the outer limit of the colony's territory; but if it did stretch continuously as far south as Montemolin and as far east as Valdecaballeros, it was of enormous extent.

Secondly, Metellinum. The colony is situated on the left bank of the Guadiana and its territory may presumably have stretched even further to the south of the river. (12) The provincial boundary in this region could not have followed the river, but must have run to the south. Since the exact extent of the territory of Metellinum cannot be ascertained, the exact location of the provincial frontier with Baetica in this zone is likewise obscure.

It is also disputed whether the Guadiana did form the provincial boundary with Baetica on the eastern side of the <u>conventus Pacensis</u>. The existence of the <u>mansio</u> called "Fines" on the Baesuris to Pax Iulia road between Serpa and Arucci to the east of the river has led to the argument that this is where the provincial boundary should be located. (13) However, the toponym "Fines" does not necessarily have to relate to an inter-provincial boundary, but could refer to the limits of a colony or <u>municipium</u>. Moreover, there is no secure evidence that Serpa was a Lusitanian town. (14) In conclusion, no clinching arguments have to date been put forward to demand modification of the view of the ancient sources, that the Guadiana formed the southern and south-eastern boundary between Lusitania and Baetica, with the exceptions being in the territories of Emerita and Metellinum already discussed. So for the purposes of this study this will be adopted as the south-eastern boundary of the province, even though it cannot definitely be established as the Roman boundary.

The eastern limit of Lusitania remains even more elusive. A hint as to its location is given by Pliny, when he is discussing the extent of Tarraconensis : (15)

> Tarraconensis autem ..... Solorio monte et Oretanis iugis Carpetanisque et Asturum a Baetica atque Lusitania distinguitur.

Mons Solorius (the modern Sierra Nevada) and the Oretani do not concern us, since they formed the boundary of Tarraconensis with Baetica, but it is the Carpetani (who occupied the central-northern Meseta) who inhabited the zone of Tarraconensis near the boundary with Lusitania. The Vettones, who bordered on the territory of the Carpetani, occupied the easternmost part of Lusitania, as the official title of the Roman provincial procurator ("procurator provinciae Lusitaniae et Vettoniae") emphasises. (16)

To supplement this rather vague orientation, Pliny's list of towns of Lusitania is useful. Caesarobriga (modern Talavera de la Reina) was the most easterly town of Lusitania. (17) To the north stretched the imposing massif of the Sierra de Gredos, which offered little prospect for settlement, and so it is not surprising that the boundary in this zone is unclear. Salmantica (modern Salamanca) was certainly a part of Lusitania, (18) but there is much doubt over Avila. The latter was clearly in Vettonian territory, but during the Augustan period belonged to Tarraconensis. (19) By the second century A.D., however, it had been transferred to Lusitania. (20)

Finally, the exact line of the provincial boundary between Avila and the Duero is unclear, but it may have followed the watershed of the Tormes and Adaja rivers.

The province itself can usefully be subdivided into three smaller units according to its Roman judicial districts, or _conventus_, established either under Augustus, Claudius or Vespasian. These were the _conventus Emeritensis_ (with its judicial centre at _Emerita_), the _conventus Pacensis_ (with its centre at Pax Iulia) and the _conventus Scallabitanus_ (with its centre at Scallabis). (21) Again the limits of these divisions are by no means certain. The boundaries of the _conventus Emeritensis_, coincided on its northern, eastern and southern sides with those of the province; its western boundary, and the eastern one of the _conventus Scallabitanus_, probably consisted of the river Coa, the river Erges and then the Tagus. The Duero formed the northern limit of the _conventus Scallabitanus_, the Tagus its southern, the Atlantic coast indisputably, one hopes, its western. As for the _conventus Pacensis_, the only problem concerns its eastern limit, which for the most part coincided with the south-eastern limit of the province, itself controversial; where the boundary was between Pacensis and Emeritensis north of where the Guadiana turns east at Badajoz is another matter for conjecture. (22)

Lusitania has had, at least since Otho's banishment thither by Nero, the reputation as being something of a backwater on the Atlantic fringes of the Roman Empire. (23) It constantly suffers by comparison with neighbouring Baetica with its high level of civilization and Roman culture from an early date, prolific as the birthplace of Roman authors, senators and eventually emperors. Its agricultural fertility, especially the fecundity of its olive groves, led to a thriving export economy. But Baetica was quite an exceptional, and thus untypical, province. The picture of Lusitania as a world apart from civilization owes much to the semi-mythological account presented by Strabo, and still has currency in most visions of the Roman Empire. (24) But Strabo used the term "Lusitania" to refer not to the Roman administrative unit, but to the ethnic area occupied by peoples generically termed "Lusitani", who essentially occupied the northern part of the <u>conventus Scallabitanus</u>. (25) He also conflates accounts of the Lusitani and the peoples of Gallaecia and Asturia. A closer examination, however, of the environmental and ethnic makeup of the Roman province of Lusitania reveals a much richer mosaic of diversity.

Ecologically the province covered many different regions, some agriculturally fertile, others windswept, barren plateaux supporting little but the characteristic <u>maquis</u> or heathland; some rich in minerals, others with good pastureland. In addition the relatively high proportion of coastline and wide river estuaries afford access to the sea, with its potential for fishing and salt production. (26)

Ethnically the province comprised a variety of peoples of differing levels of civilization. (27) The Algarve and southern Alentejo had been inhabited by the Conii, or Cynetes, who had settled in proto-urbanized communities and had developed an alphabet and script, before they were overwhelmed by Celtic invaders from the central Spanish Meseta in the fifth or fourth century B.C. (28) These Celts occupied the whole of the Alentejo and settled in characteristic hill-forts (or "castros"). But there were also immigrants from Turdetania (part of Roman Baetica), Turduli, in this region, who settled in proto-urban communities : Balsa (Tavira), Ossonoba (Faro), Myrtilis (Mertola), Pax Iulia (Beja), Salacia (Alcácer do Sal) and Caetobriga (Setúbal). Emerita was also located in Turdulian territory. Most of these communities had minted their own copper coinage in the pre-Augustan period, another mark of their relatively advanced civilization. (29)

The territory to the north of the Tagus in modern Portugal was occupied by the Lusitani, probably a generic term covering a multiplicity of smaller people (<u>populi</u>), who also settled in hill-top enclosures. The river valleys and coastal areas, however, of this region, being more open to cultural influences from outside, saw a greater development of proto-urban communities in the pre-Augustan age : for example, Talabriga (? Marnel), Aeminium (Coimbra), Conimbriga (Condeixa-a-Velha), Collippo (S. Sebastião do Freixo), Eburobrittium

FIGURE 2.2 <u>NATIVE PEOPLE OF LUSITANIA</u>

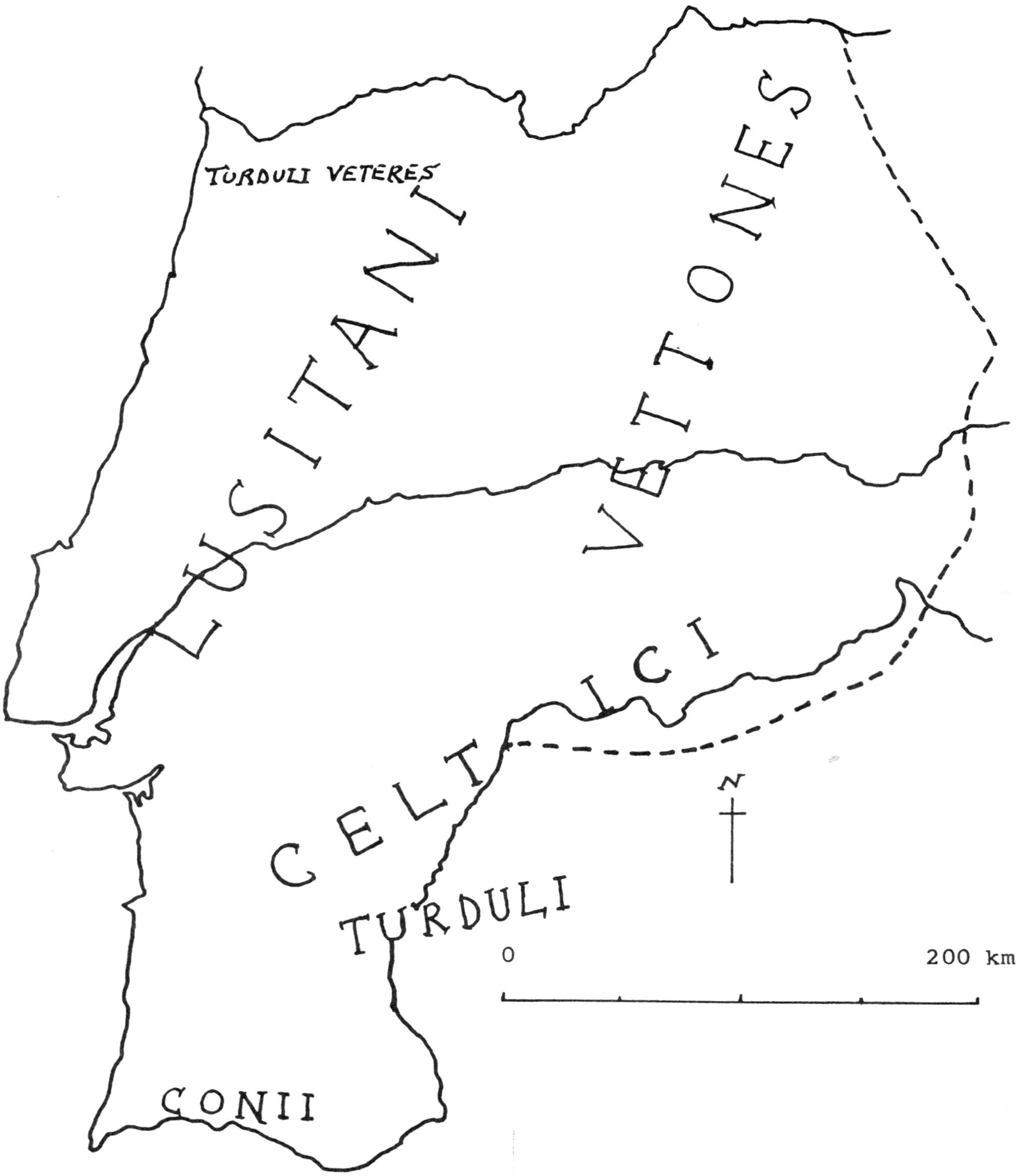

(Amoreira de Obidos), Sellium (Tomar) and Olisipo (Lisbon). (30) The territory to the north of the Tagus in Spain was occupied by the Vettones, whose material culture was essentially Celtic, who settled in similar types of hill-fort to those of the Lusitani and whose economy was essentially centred on stock-raising, as opposed to agriculture. (31)

In broad terms there was a marked contrast between the Celticized interior of the province, marked by a "castro culture" and those regions more accessible to external influences, and more developed societies, notably the Phoenico-Punic and Greek colonies along the southern coast of Spain, and the relatively advanced societies of the area that came to comprise Roman Baetica. It is no coincidence that all the communities of Lusitania of privileged status (i.e. colonies or *municipia*) under Augustus had their origins as proto-urban settlements in the later Iron Age in areas in contact with the outside world. (32)

A crude index to the agriculturally promising regions of the province is provided by the overall distribution of Roman rural villa sites, since most Roman villas were the centres of agricultural estates. (33) It must be stressed that the villa was not the only means of exploiting the land. Other possibilities included small-scale scattered farmsteads of a more modest nature, built of wood, thus leaving little trace in the archaeological record; smaller rural nucleated settlements (such as *vici*) that acted as centres of agricultural communities; or even a system whereby farmers resided in towns and travelled to their fields daily to work. Moreover, the villa system may only have taken root in the more Romanized parts of the province. Taking these drawbacks into account, the spread of villas at least provides some sort of impression as to where the richer agricultural zones may have been, even if what is charted may only be considered the minimal extent of the agricultural regions.

FIGURE 2.3 ROMAN VILLAS IN LUSITANIA

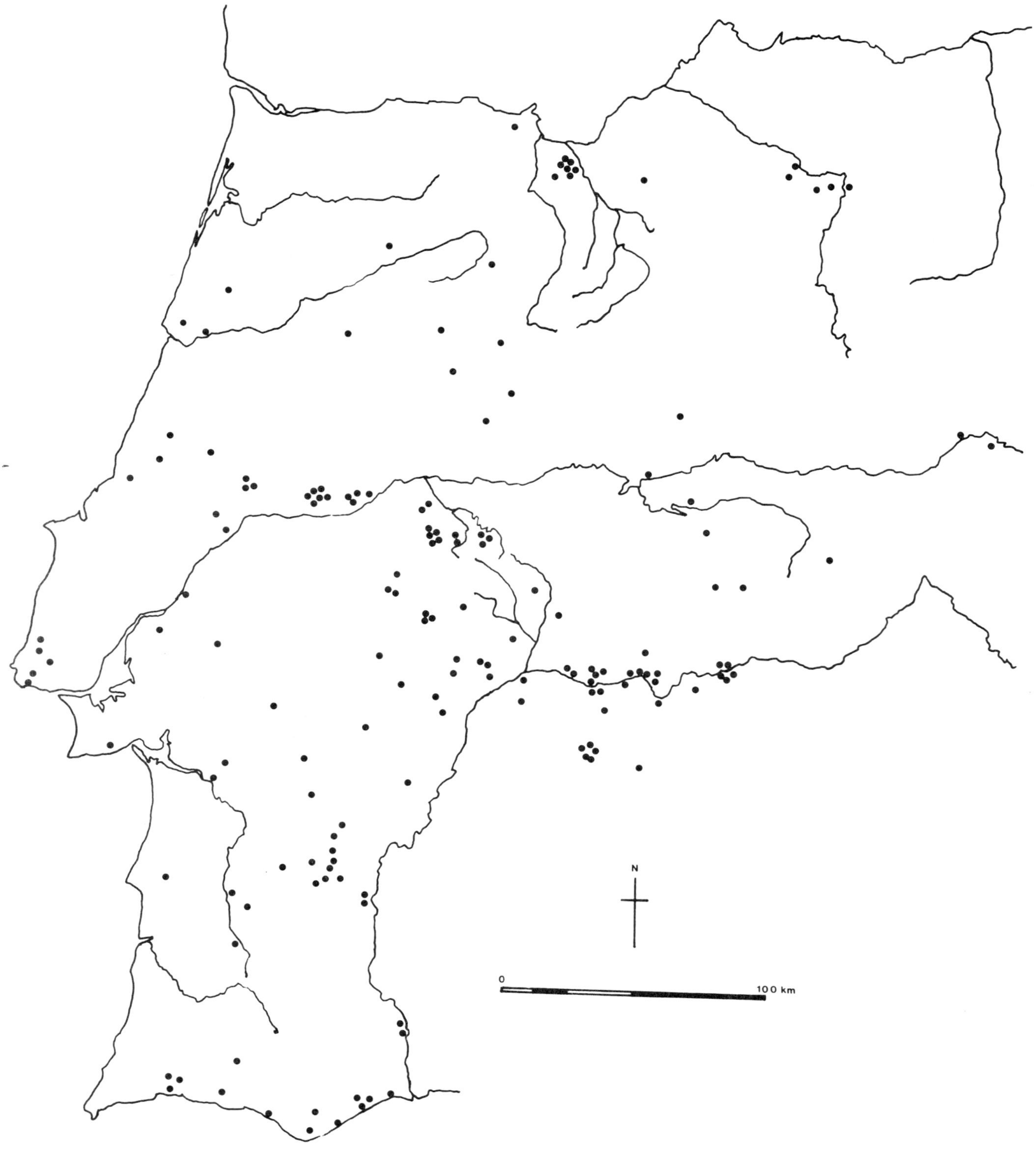

All sites catalogued in J.-G. Gorges, Les villas hispano-romaines, Paris, 1979 are shown in Figure 2.3 with the following additions :

(a) Conventus Pacensis

1. Monte da Chamine, Ferreira do Alentejo (Beja) : Inf. arq. 1982-83, 41-2.
2. Figueira de Metade, Herdade de Conqueiros, Alvalade-Sado (Santiago do Cacem) : Inf. arq., 1979, 79-82.
3. Monte da Romeira, Viana da Alentejo (Évora) : Inf. arq., 1982-83, 13-16.
4. Reguengos de Monsaraz, Azraieira (Évora) : kind information of Prof. Alarcão.
5. Monte da Parreira, Benavente : Inf. arq., 1980, 14-15; Arqueologia (Porto), 3, 1981, 132.
6. Monte da Herdade dos Pombais, Marvão (Portalegre) : Inf. arq., 1982-83, 32, 96-97.
7. Quinta do Pião, Alter do Chão (Portalegre) : Actas das III Jornadas Arq., I, 278-9
8. Monte de Alfarofia, Caia, Elvas : Inf. Arq., 1977-78, 80.

(b) Conventus Emeritensis

1. Torreaguila, Barbano, Montijo (Mérida) : kind info. of J.M. Alvarez Martinez.
2. El Pesquero, Pueblonuevo del Guadiana, Talavera (Badajoz): XVII Congreso Nac. de Arq. Logroño 1983, Zaragoza, 1985, 487; Arqueologia 83, No. 10.2; Arqueologia 82, No. 48.
3. El Lobo, San Roque (Badajoz) : XV Congr. Nac. de Arq., Lugo 1977, Zaragoza, 1979, 908.
4. Dehesa Boyal, Berzocana (Cáceres) : Arqueologia 79, No. 89.
5. El Soto, Calvarrasa de Abajo (Salamanca) : Zephyrus, 32-33, 1981, 217-25.

(c) Conventus Scallabitanus

1. Alto do Cidreira, Carrascal de Alvide, Alcabideche (Cascais) : Inf. arq., 1980, 39-40; Inf. arq., 1982-83, 82-3.
2. Casais Mouros, Selão, Areia (Cascais) : J. d'Encarnação, Notas sobre alguns vestigios romanos no concelho de Cascais, Estoril, 1968, 10.
3. Brandoa, Sintra/Amadora (Lisbon) : Inf. arq., 1981, 22.
4. Moinho do Meio, Abrantes : Arqueologia (Porto), 4, 1981, 124-30.
5. Vila Nova de Ourem, Olival : kind info. of Prof. Alarcão.
6. S. Pedro de Caldelas, Madalena (Tomar) : Inf. arq., 1979, 72; 1980, 53-54; 1981, 96-97.
7. Ferro, Covilhã : M.A.G. Carvalho Rodrigues, Estudos de Castelo Branco, 50, 1974, 67.
8. Quinta do Costa, Guimarães de Tavares, Chas de Tavares, Mangualde (Viseu) : Inf. arq., 1982-83, 144-45.
9. Vila Verde de Ficalho (centre) : Inf. arq., 1982-83, 32.
10. Quinta de Pero Martins, Guarda : M. Maia, AP, iii, 7-9, 1974-77, 210-211.
11. Pedregais, Guarda : ibid., 212.
12. Vale de Olmos, Vermiosa, Guarda : ibid. 212.

(a) Conventus Pacensis

The territory around Pax Iulia (modern Beja) has revealed a relatively high density of Roman villas. The Lower Alentejo has always been one of the most productive wheat growing areas of Portugal and so it is likely that this was one of the more fertile regions of the Roman province. Ebora (modern Evora) is now surrounded by relatively fertile plains, but few traces of Roman rural settlement have to date been found in the territory of this Roman municipium. This is the one area of the province from where Roman senatorial families are known : the Q. Iulii Maximi owned the villa at Nossa Senhora da Tourega. (34) The Algarve coastal strip, although narrow, is relatively fertile, especially now for fruit crops. Elsewhere in the conventus the terrain is less promising.

Immediately to the north of the Algarve rise several Serras, to the north of which lies the vast plateau of the Alentejo. This is basically a wide expanse of heathland, now supporting significant olive groves, and still more plantations of evergreen oak, notably the cork oak. The western coastal regions are marked by sandy soils, which support little but pine woods.

(b) Conventus Emeritensis

Another fertile area of the province is the so-called Tierra de Barros, located in what once was the territory of the provincial capital, Augusta Emerita. It is in just this area that centuriated land has been discovered, i.e. land allotted to the Roman colonists. (35) Conditions are also reasonably favourable around Caesarobriga (modern Talavera de la Reina) at the extreme south-east of the conventus, and around Salmantica (modern Salamanca) in the Castilian plains. But although soil conditions are promising, the climate is less so, consisting of "tres meses de invierno, nueve de infierno" ("three months of winter and nine of Hell") according to a local proverb of Extremadura. It is no coincidence that this region of Spain saw the highest level of emigration to the new territories in the Americas : Cortes himself came from Medellin, the Pizarro brothers from Trujillo. (36) The interfluvial zone between the Guadiana and the Tagus consists of a vast tableland of rocky soils which support little but heath. North of the Tagus rise the central Sierras (the Sierra de Gredos, de Gata and de Francia), before the terrain levels out again into the Castilian plateau either side of the Duero.

Much then of this *conventus* could support little agriculture; stock-raising always has been more dominant in the economic pattern of the area, except in the regions already singled out.

(c) Conventus Scallabitanus

Favourable agricultural conditions prevail in the valley of the Tagus between Olisipo and Scallabis, for agriculture, as well as for vines and olives. To the north the valleys of the Vouga and Mondego provide some good alluvial soil, and agriculture here is favoured by the more temperate Atlantic climate. However, in the interior and between these river valleys the landscape is broken up by inhospitable Serras, which provide little prospect of successful agriculture. Even now these regions support very low densities of settlement. In the south-east of the *conventus* (the modern district of Castelo Branco) the landscape is scarred by numerous tributaries of the Tagus and small Serras, where agriculture is difficult on anything but a small scale, but which is blessed by widespread mineral resources.

It will be important to discuss the development of "industrial" processes against this background. To what extent were they "subsidiary" activities located in areas of agricultural fertility ? Or did they form the dominant element within the local economy ? Who controlled the production, landowners or independent operators ? In short, what effect did these industries have on the regional economy of the Roman province of Lusitania ?

NOTES : CHAPTER TWO

1. For the division see E. Albertini, Les divisions administratives de l'Espagne romaine, Paris, 1923, ch. 3 : dating it between 12 and 2 B.C., probably between 7 and 2 B.C.; R. Syme, 'The conquest of north-west Spain', Legio VII Gemina, León, 1970, 83-107 = Roman Papers (ed. E. Badian), II, Oxford, 1979, 825-54 prefers the earlier date of 16 - 13 B.C. For Mérida see I.A. Richmond, 'The first years of Emerita Augusta', AJ, 87, 1930, 98-116. J.A. Saenz de Buruaga, 'La fundición de Augusta Emerita' in Augusta Emerita : Actas del Bimilenario, Madrid, 1976, 19-32. On the military organization : P. le Roux, L'armée romaine et l'organisation des provinces ibériques d'Auguste à l'invasion de 409 (= Publ. du Centre Pierre Paris, 8), Paris, 1982. On boundary changes : F. Braun, Die Entwicklung der spanischen Provinzialgrenzen im römische Zeit, Berlin, 1909.

2. Strabo 3.4.20 (166); Plin. NH 4.21.113; Martianus of Heraclea, 10, 11, 13 : for text see M. Pastor Munoz, 'La peninsula iberica en Marciano de Heraclea', Hisp. Ant., 8, 1978, 89-128.

3. Plin. NH 3.1.6; 4.22.115; Pomponius Mela 2.87; 3.6; Martianus of Heraclea 2, 8, 13.

4. The problems have been stated and discussed by L. Garcia Iglesias, 'El Guadiana y los limites comunes de Betica y Lusitania', Hisp. Ant., 2, 1972, 165-77.

5. For a full discussion see R. Wiegels, 'Zum Territorium des augusteischen Kolonie Emerita', MDAI(M), 17, 1976, 258-84.

6. De Controv. Agr. 51 f. (Lachmann) = 44 (Thulin)

7. J.-G. Gorges, 'Centuriation et organisation du territoire : notes préliminaires sur l'exemple de Mérida' in P.A. Fevrier & P. Leveau (ed.), *Villes et campagnes dans l'Empire romain*, Aix-en-Provence, 1982, 101-110; P. Sillières, 'Centuriation et voie romaine au sud de Mérida : contribution à la délimination de la Bétique et de la Lusitanie', *MCV*, 18, 1982, 437-48.

8. J.-G. Gorges, art. cit. (n.7), 104 for Sierra de Zafra; watersheds were often used by the Romans for fixing boundaries : Hyginus, *De Condic. Agror*. 114 (Lachmann) = 74 (Thulin); *De Gener. Controvers*. 134 (Lachmann) = 97 (Thulin).

9. Montemolin : *HAEp* 1483; Valdecaballeros : *CIL* II 656 = *ILS* 5972.

10. There are parallels for a community owning land away from its urban centre : Arpinum drew revenue from land owned in Cisalpine Gaul : Cic. *ad Fam.* 13.11.1; *CIL* V 7749, line 25. See in general F.F. Abbott & A.C. Johnson, *Municipal Administration in the Roman Empire*, Princeton, 1926, 138; A.H.M. Jones, *The Roman Economy* (ed. P.A. Brunt), Oxford, 1974, 28; W. Liebenam, *Städteverwaltung im römische Kaiserreiche*, Leipzig, 1900, 6-10 (with further examples).

11. So J.-G. Gorges, art. cit. (n. 7), 103-104.

12. A. Garcia y Bellido, 'Las colonias romanas de la provincia de Lusitania', *Arqueologia e Historia*, 8, 1958, 13-23, esp. 13-14.

13. E. Albertini, op.cit. (n. 1), 39-40; R. Thouvenot, *Essai sur la province romaine de Bétique*, Paris, 1940, 165.

14. See A. Tovar, *Iberische Landeskunde II. Die Völker und die Städte des antiken Hispanien. 2. Lusitanien*, Baden-Baden, 1976, 170.

15. Plin. *NH* 3.1.6.

16. *CIL* II 1267; cf. *CIL* II 485 (*tabularius provinc. Lusitaniae et Vettoniae*)

17. Plin. *NH* 4.22.118.

18. A. Tovar, op.cit. (n. 14), 245-246.

19. Note the characteristically Vettonian hill-fort at Las Cogotas, Cardeñosa, 12 km from Avila : J. Cabré Aguilo, *Excavaciones de las Cogotas, Cardeñosa (Avila). I. El Castro* (= *MemJSEA*, 110), Madrid, 1930; id., *II. La necropolis* (= *MemJSEA*, 120), Madrid, 1932. Plin. *NH* 3.3.19, however, does include the Vettones among the people of Hispania Citerior (= Tarraconensis).

20. Ptol. 2.5.7; F. Braun, op. cit. (n. 1), 117; E. Albertini, op.cit. (n. 24), 115-116 (Avila as part of Lusitania in terms of Visigothic bishoprics).

21. Augustus : N. Mackie, *Local Administration in Roman Spain A.D. 14 - 212* (B.A.R. Int. Ser., 172), Oxford, 1983, 8; Claudius : E. Albertini, op.cit. (n. 1), 54-67; Vespasian : R. Etienne, *Le culte imperial dans la péninsule ibérique d'Auguste à Dioclétian*, Paris, 1958, 185-9.

22. Plin. *NH* 4.21.117

23. Tac. *Hist*. 1.21; cfr. 1.13

24. On Strabo as a source see A.D. Momigliano, *Alien Wisdom : the limits of Hellenization*, London, 1975, 65-8. For the modern picture : J.M. Blazquez, *La Romanizacion II : La sociedad y la economia en la Hispania romana* (= Ciclos y temas de la historia de España), Madrid, 1975, esp. 188 : "Al comienzo del Principado ...... la romanización habia avanzado poco en Lusitania".

25. 3.3.3 (152). He also organized his discussion of Asia Minor according to ethnic areas in books XI to XVI.

26. For a full account see Royal Naval Intelligence Division, Spain and Portugal (Geographical Handbook Series), 3 vols., London, 1941; see also J. C. Edmondson, Economy and Society of Roman Lusitania under the Principate (unpubl. Ph.D. thesis), University of Cambridge, 1984, ch. 1 : "Historical Geography : the limits of the possible".

27. For a general introduction see J. de Alarcão, Portugal romano (3rd ed.), Lisbon, 1983, 17-21.

28. M. Maia, 'Povos do sul de Portugal nas fontes classicas - os Conios' in Actas do IV CNA Faro 1980 (forthcoming); M. Almagro Basch, Las estelas decoradas del suroeste peninsular (= Bibliotheca Praehistorica Hispana, VIII), Madrid, 1966, esp. 210-211; L. Coelho, 'Inscrições da necropole proto-historica da herdade do Pego - Ourique', AP, iii, 5, 1971, 167-180.

29. On the Celtici see M. Maia, 'Povos do sul de Portugal nas fontes classicas - Celtici e Turduli', Clio, 2, 1980, 67-70; for the introduction of Celtic ceramics to the area : J.M. Arnaud & T.J. Gamito, 'Ceramicas estampilhadas da idade do ferro do sul de Portugal I - Cabeça de Vaiamonte, Monforte', AP, iii, 7-9, 1974-1977, 165-202; for coins see A. Vives y Escudero, La moneda hispanica, III, Madrid, 1924, 81 (Baesuris = Esuri), 114 (Ossonoba), 85 (Silves = Cilpe), 24-27 (Salacia), 90-91 (Myrtilis).

30. J. de Alarcão, op.cit. (n. 27), 18-21.

31. J.M. Roldan Hervas, 'Fuentes antiguas para el estudio de los Vettones', Zephyrus, 19-20, 1968-1969, 73-106.

32. For privileged communities see Plin. NH 4.22.117. For evidence of Iron Age settlement at Roman colonies : Emerita : J.M. Alvarez Martinez,

'Consideraciones sobre la Mérida prerromana', *REstExtr*, 40, 1984, 101-9; Pax Iulia : F. Nunes *Ribeiro*, 'Prehistoria e a origem de Beja', *Arq. de Beja*, 17, 1960, 3-113; Scallabis : M.M. Arruda, 'Alcaçova de Santarém. Relatorio dos trabalhos arqueológicos de 1984', *Clio/Arq.*, 1, 1983-4, 217-224; Norba Caesarina : C. Callejo Serrano, 'La arqueologia de Norba Caesarina', *AEA*, 41, 1968, 121-49, esp. 121; Metellinum : M. Almagro Gorbea, 'La necropolis de Medellin', *NAH*, 16, 1971, 161-202; at *municipia* : Olisipo : A. Vieira da Silva, *Epigrafia de Olisipo : subsidios para a historia de Lisboa romana*, Lisbon, 1944, 40-41; Salacia : C. Tavares da Silva, J. Soares, C. de Melo Beirão, L. Ferrer Dias & A. Coelho-Soares, 'Escavações arqueologicas no Castelo de Alcácer do Sal (campanha de 1979)', *Setúbal Arq.*, 6-7, 1980-81, 149-218, esp. 181-7; Ebora : A. García y Bellido, 'El recinto mural romano de Evora', *Conimbriga*, 10, 1971, 85-92, esp. 87; Myrtilis : L.F. Delgado Alves, 'Aspectos de arqueologia em Myrtilis', *Arq. de Beja*, 13, 1956, 41-45.

33. J.-G. Gorges, *Les villas hispano-romaines : inventaire et problématique archéologiques*, Paris, 1979.

34. R. Étienne, 'Sénateurs originaires de la province de Lusitanie' in S. Panciera (ed.), *Epigrafia e Ordine Senatorio* (= *Tituli*, 5), Rome, 1982, 521-529.

35. See note 7 (supra).

36. R. Syme, *Colonial Elites : Rome, Spain and the Americas*, Oxford, 1958, 29.

CHAPTER THREE

THE MINES OF LUSITANIA

i. Introduction

In the Roman period metals of all kinds were needed on a large scale for a wide variety of uses : for coinage, building materials, arms and armour, tools and utensils, as well as luxury items such as jewelry and furniture. (1) As Romanization spread throughout the Empire, the demand for metals substantially increased. The monetization of the economy, the dramatic increase in urbanization (especially in the western part of the Empire) and the presence of a large standing army on the frontiers all stimulated the need for a regular supply of many types of metal. As the Roman Empire expanded, new areas of supply came within the sphere of exploitation of the Roman state, notably the Iberian peninsula, Gaul, Britain, Dacia, Thrace and Asia Minor. As a result of this, the Roman state derived enormous revenue from the leasing out of contracts to work imperially owned mines. (2)

That the Iberian peninsula constituted one of the main sources of metals - perhaps the most important source - is not in doubt. (3) It has been commonplace for both ancient and modern authors to eulogize this mineral wealth. Strabo, for example, praised the whole peninsula, singling out Turdetania for special mention. (4) Diodorus Siculus (probably reproducing the account of Posidonius) concentrated on the silver mines of Carthago Nova, but also mentioned the tin to be found elsewhere in the peninsula. (5)

Recent scholarly attention has concentrated on the mines of two regions of the peninsula : the gold mines of north-west Spain (6) and the silver and copper mines of the Sierra Morena, centred on the mines of Rio Tinto. (7) Alternatively, attempts have been made to synthesise the literary evidence for Roman mining in the peninsula as a whole, with occasional reference to archaeological material. (8)

The literary evidence, scant though it is, makes it quite clear that the province of Lusitania should not be excluded from these discussions. Strabo talks of the auriferous rivers of the province : the Duero, the Vouga, the Mondego and the Tagus, (9) he refers to the great abundance of gold, silver and other metals that existed between the Tagus and the Artabrians in the far north of the peninsula (10) and the ore-carrying mountains that stretched between the Tagus and the Guadiana rivers (11) and, finally, he mentions the tin that was extracted in the country beyond the territory of the Lusitanians. (12) Here he is most probably referring to the important stanniferous zone in the north-east of Lusitania, which extended well into Hispania Tarraconensis.

The elder Pliny also talks of the alluvial gold of the Tagus (13) and reports that the annual gold production of Asturia, Callaecia and Lusitania amounted to some 20,000 pounds (with Asturia the largest contributor). (14) This quantity of gold was equivalent to 84,000,000 sesterces at contemporary prices, an enormous sum. (15) Pliny also confirms the presence of tin (plumbum candidum) in the province, which was obtained by alluvial washing or by extraction from certain kinds of gold mine. (16) Finally he talks of the exploitation of lead in the Iberian peninsula (17) and provides some clue as to its exploitation in Lusitania when he stresses that the cognomen of a Lusitanian people, the Medubrigenses, was "Plumbari". (18)

There are problems, however, with accounts of Roman mining based too exclusively on literary evidence. The extraction of gold from the Tagus is a case in point. A recent article has collected some forty-five references to "gold-bearing Tagus" scattered in the works of some twenty-two Latin writers ranging in time from Catullus to Boethius. (19) Gold, as we shall see, was certainly extracted from the Tagus in Roman times, since various alluvial terraces have been discovered. But references to this gold in poems written in the late fourth/early fifth century A.D. by poets such as Claudian, Prudentius and Rutilianus Namatianus are scarcely sufficient to prove that gold was still being extracted from the river in this period, since "aurifer Tagus" had clearly longsince become a literary topos. (20)

It is only by reference to archaeological material that the literary evidence can be verified. Archaeological work can help to locate more precisely the areas of Roman mining interest, can provide more exact information on the minerals extracted and can also give some indication of the periods of operation of a particular mine. There are, however, certain limitations to the archaeological evidence for mining, which need outlining immediately :

1. Our evidence is distinctly patchy : some areas have received much more archaeological attention than others and thus the relative importance of mining in one district as opposed to another is often very difficult to assess. Our picture of Lusitanian mining must, therefore, be taken as provisional.

2. Quite a few Roman mines owe their discovery merely to the fact that they were reopened and reworked in the nineteenth and twentieth century. Those Roman mines that were exhausted in the Roman period stand little chance of discovery. This factor means that we are more likely to have knowledge of only the larger Roman mines and thus an imbalanced picture of the pattern of mining in any one area, in which a few large-scale mines loom large to the exclusion of smaller ones.

3. Furthermore, alluvial workings are much more difficult to locate archaeologically than underground mines proper. This is again likely to give rise to an imbalanced picture of the pattern of Roman mining in a particular area.

4. Although we are at first thankful that the recent reactivation of a mine has revealed traces of its Roman exploitation, the blessing can be somewhat mixed. For subsequent exploitation tends to obliterate the traces of a mine's former working. Especially with the advent of modern technology, the rate of mining is now substantially greater than it was, say, fifty years ago; and the faster the rate of mining, the less chance there is of discovering ancient features.

5. Recent mining will also have removed crucial evidence for the scale of the Roman operations. Roman slag heaps have been located and investigated at many major mining sites, but they may well be a deluding guide to the scale of Roman mining. For slag has often been resmelted (either in antiquity or more recently), thus removing the evidence for earlier working. A clause of a Roman mining regulation from Vipasca (Aljustrel) deals with the resmelting of copper and silver slag, while at

the silver mines of Laurion in Attika fifth and fourth century B.C. slag was resmelted in the later Roman Empire. (21)

6. Finds of clearly datable Roman artefacts are very rare in mines themselves. We almost always have to rely on pottery or coins found in a connected settlement or burial ground. But it does not always necessarily follow that a mine was operational for exactly the same period as a nearby settlement.

7. Finally, at many sites the mere fact that a mine had been worked in the Roman period has often been noted, but no further investigation has been carried out. It is often, therefore, difficult to assess the extent of the Roman working of the site and impossible to be precise over the exact periods of its operation.

Despite these problems over the evidence, a catalogue of known Lusitanian mines has been prepared (Appendix I), provisional though it be, to give some impression of the general areas of Roman mining in the province. Concerning the type of metal extracted at any one mine, it is unwise to be too dogmatic, since a variety of metals could be obtained from a single mineral source. So Strabo remarks that in Spain copper-mines were also termed gold-mines, since gold was previously the main metal extracted. (22) Lead could be obtained as a by-product of the smelting of silver : lead ingots survive bearing the marks EX ARG. or EX ARGENT.; (23) others have been recovered from the mines of Carthago Nova, where silver was the main product. (24) Copper and silver were the main concerns at Vipasca (P 33), but gold and even iron may also have been extracted. (25) At Rio Tinto in Baetica gold, silver, copper, lead and iron were all contained in the ore body. (26)

ii. The Geographical Distribution of Mines

Not only from geological work on the ore bodies, but also from an assessment of the archaeological evidence for Roman mining in the province, it is clear that there were, broadly speaking, three main areas of mining in Lusitania, concerned with the extraction of four major minerals, gold, silver, copper and tin. (27)

Silver and copper were exploited in the southern part of the province at the westernmost extreme of the Iberian Pyrite Belt (which stretches some 250 kilometres over south-west Spain and south Portugal and in the Roman period contained the important Baetican mines at Rio Tinto, Sotiel Coronada, Tharsis and São Domingos). Secondly, gold was obtained mostly from the conventus Scallabitanus from alluvial workings along the Tagus and its tributaries (especially the Zêzere and the Erges), the Mondego and the Douro. In addition gold mines were exploited in the Serra da Lousã. These strains of gold extended into north Portugal, Galicia and Asturia, where a heavy concentration of Roman gold mines has been discovered. (28) Thirdly, tin was mined in the north-eastern part of the conventus Scallabitanus and in the north of the conventus Emeritensis - a mining zone that extended into the modern Spanish provinces of Orense, Zamora and Pontevedra in Roman Tarraconensis. (29) This overall distribution of mining sites is shown in Figure 3.1, and summarized in Table 3.2.

FIGURE 3.1 THE MINES OF LUSITANIA : GENERAL DISTRIBUTION

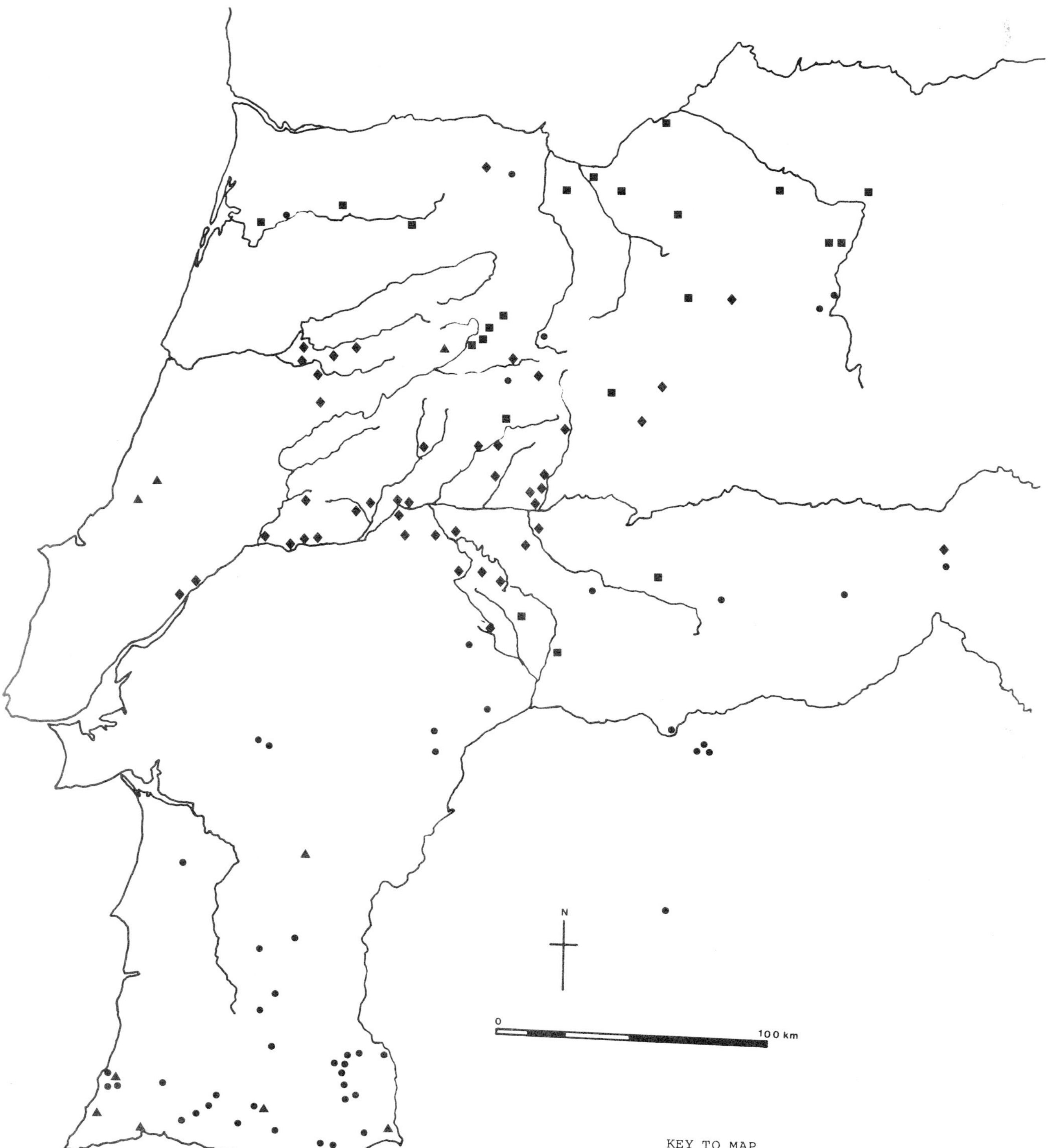

KEY TO MAP

- ● Silver/Lead/Copper
- ◆ Gold
- ■ Tin
- ▲ Iron/Manganese

TABLE 3.2. DISTRIBUTION OF MINES ACCORDING TO <u>CONVENTUS</u> AND MAIN METAL EXTRACTED

(for details see Appendix I)

CONVENTUS	Cu/Ag/Pb	Au	Sn	Fe/Man	TOTAL
Pacensis	35	6	1	6	48
Emeritensis	11	10	13	-	34
Scallabitanus	3	29	9	3	44
TOTAL	49	45	23	9	126

Even though it is possible to gain an overall picture from such distribution maps, because of the lack of detailed archaeological investigation it is often very difficult to assess the scale of Roman operations at any one site. It is, however, clear that there were different modes of exploitation, which depended on the sheer extent of the mineral deposits. On the one hand, there were the large mining districts, where the extraction of metals formed the major economic activity of the area. But there were also much smaller concerns, where minerals were discovered on private land and a few shafts sunk by, or with the permission of, the local landowner.

FIGURE 3.3 MINES OF THE CONVENTUS PACENSIS

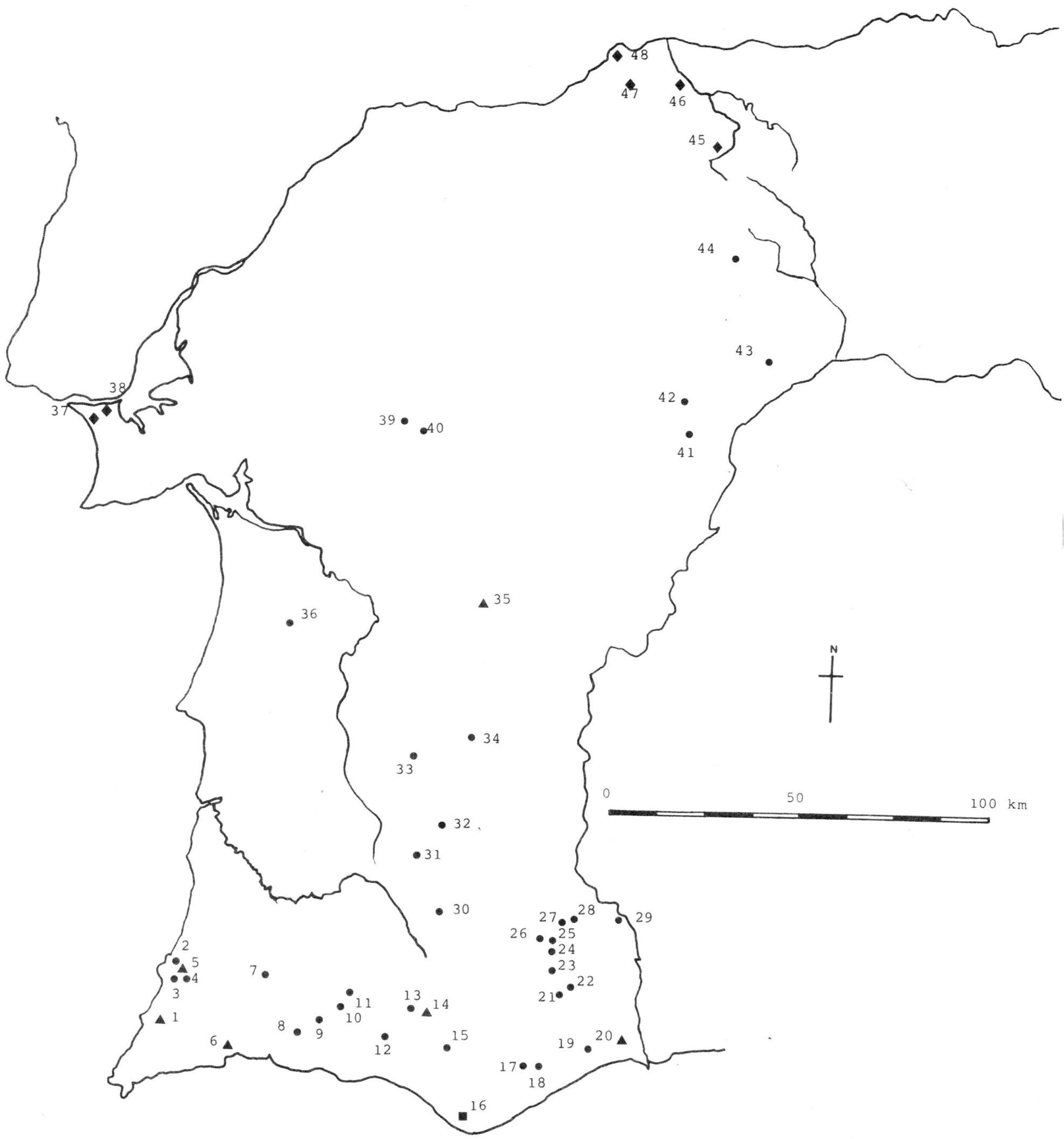

KEY TO MAP

● Silver/Lead/Copper ■ Tin

◆ Gold ▲ Iron/Manganese

n.b. numbers refer to Catalogue of Mines (Appendix I)

FIGURE 3.4 MINES OF THE CONVENTUS EMERITENSIS

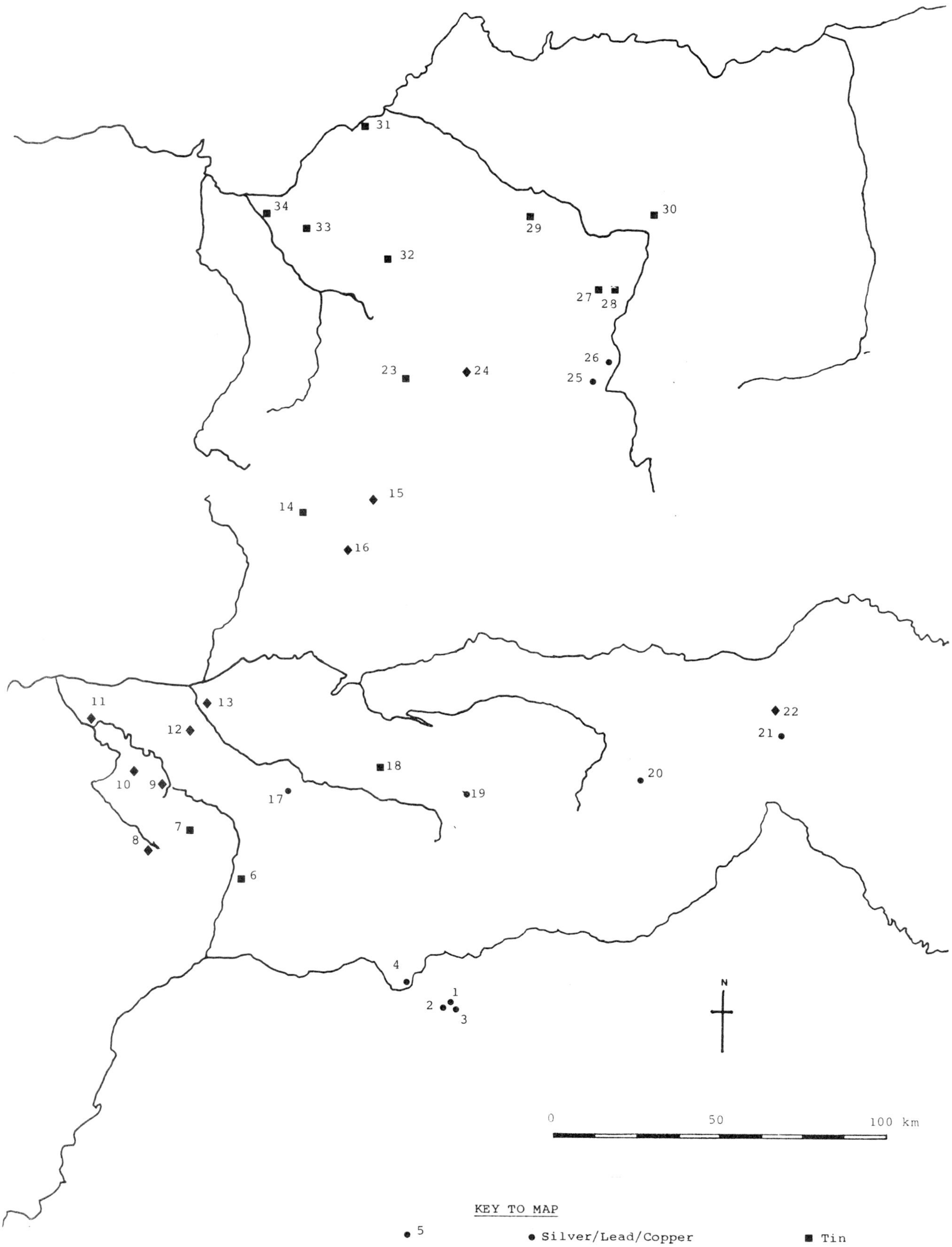

KEY TO MAP

● Silver/Lead/Copper

■ Tin

◆ Gold

▲ Iron/Manganese

n.b. numbers refer to Catalogue of Mines (Appendix I)

FIGURE 3.5 MINES OF THE CONVENTUS SCALLABITANUS

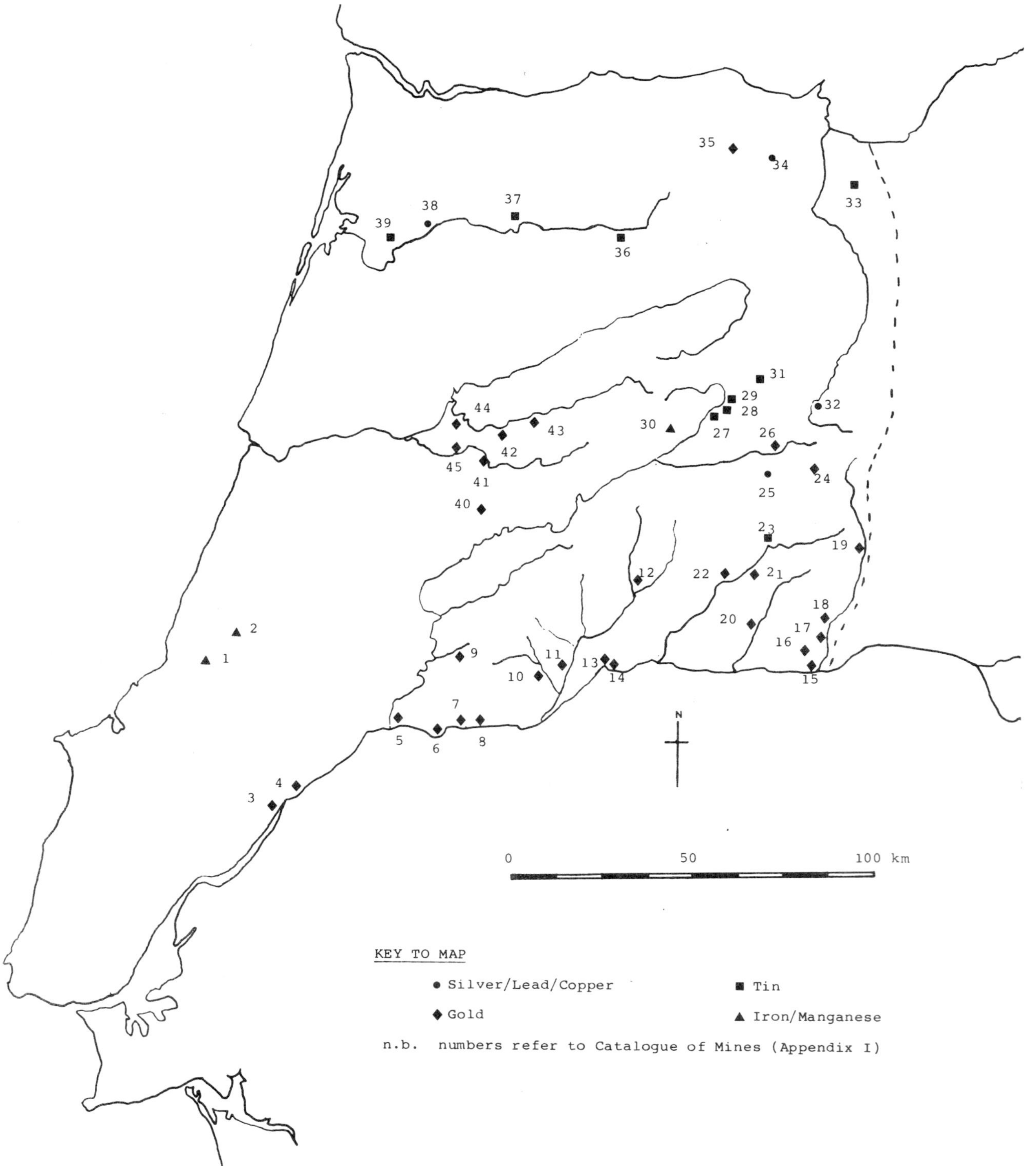

KEY TO MAP

• Silver/Lead/Copper

■ Tin

♦ Gold

▲ Iron/Manganese

n.b. numbers refer to Catalogue of Mines (Appendix I)

The most important mining district in the province was that at Vipasca (P 33), which will receive fuller attention below. Just to the east of the Anas, the São Domingos mines were also large-scale; (30) but if the Anas is taken as the provincial boundary, then the mining district belongs not to Lusitania, but to Baetica. (31) The sheer quantity of surviving slag from the copper mines at Serra da Caveira (P 36) suggests that these mines also must have been extensively exploited. (32)

Of the gold workings along the Tagus and its tributaries, there seem to have been areas of fairly intensive exploitation, perhaps suggesting a mining district. So, for example, at Constancia (S 05) three terraces for the extraction of gold stretched for three kilometres along the Tagus, as well as another terrace along the Zêzere; around Milreu (S 09) there were three areas of workings; along the Rio Ponsul (S 21) workings extended for some eight kilometres; at Meimoa (S 26) there were workings over an area of five kilometres, concentrated around Covão do Urso and La Presa.

This concentration of gold resources stretched further east into the conventus Emeritensis (mines E 8 - 13, 15). Major sites existed at Plasenzuela (E 19), where some three thousand ancient shafts were visible at the end of the nineteenth century, and at Nava del Ricomalillo (E 22), with its three zones of gold mines at La Rica, Ingeniera and Eugenia. Although most of the Roman gold mines of Roman Spain were located in the north-west of the peninsula, and most recent attention has focused on production there, the density of sites along the Tagus and the scale of operations at Plasenzuela and Nava del Ricomalillo clearly indicate that they should not be excluded from any account of Roman gold mining in the Iberian peninsula. (33) The evidence is too scant to allow any quantification along the lines of that attempted recently for the north-west, but these mines clearly formed one of the major economic assets that accrued to the Roman state after the annexation of Lusitania as a province.

Another area with a significant concentration of mineral resources was that now covered by the modern Portuguese concelho of Belmonte, with tin the major metal involved. To date, at least three Roman alluvial workings have been discovered. (S 27, 28, 29).

The scale of mining in these districts was clearly much greater than that at some of the mines in the Algarve, where, as a general rule, small numbers of shafts are reported at Roman mining sites : for example, four at Adualho (P 06), two at Valle de Pegas (P 12), single shafts at Margalho e Penedo (P 02), Cisterna dos Cães (P 08), Odeleite (P 19). Small-scale open-face workings are reported at Herdade da Corte do Sobro (P 05) and Forra Merendas, Vaqueiros (P 22).

It is, therefore, clear that there was a wide difference in the scale of the mining operations in the province, but was there also a difference in their legal and administrative organization ? The lack of precise evidence on this point excludes anything but a few general comments.

iii. The Administration of Lusitanian Mines

It is generally held that gold and silver mines had by the early Principate passed into Imperial control. (34) Some might go further and suggest that all mines of an imperial province (such as Lusitania) were ipso facto imperial property and were, therefore, administered by the Emperor. But it is rather an extreme position to hold that all mines were controlled directly by the Roman administration. Gold mines were all probably under imperial control by the Augustan period, given the value of the resource. But under the early Principate silver and other mines were still in private hands. (35) It is dangerous to extrapolate too indiscriminately from the legal texts from Vipasca (P 33), since the mines there constituted part of an imperial estate, supervised by an imperial procurator who let out contracts for the exploitation of the minerals to lessees. This arrangement can be directly paralleled in the administration of agriculture on imperial estates. (36) On private land, however, the ownership of the metalliferous subsoil went hand in hand with the ownership of the land itself. (37) A text of the Digest shows that a usufructuary was free to exploit a source of minerals, as long as he did no damage to crops. (38) There was, however, a tendancy over time for more and more land to pass into the hands of the emperor : forcible acquisition after civil wars and especially bequests of land to the emperor on the death of a local landowner represent two of the commonest mechanisms.

Just as Tiberius acquired control over the gold mines of Sex. Marius in Baetica, it may have been possible for later emperors to take over the richer mining concerns in Lusitania. (39) But it seems a priori unlikely that they would have shown much interest in very small sources of mineral, which would best have been exploited on a small-scale by the landowner concerned. The administrative effort required would not have been worthwhile in terms of possible revenue gained.

There is also uncertainty how metals, once extracted, were exchanged. For imperially owned mines the Vipasca texts reveal that the fisc received half the ore extracted, although the contractor could buy out the half belong to the fisc for the sum of 4,000 sesterces. (40) But how did the lessees dispose of the other half ?

Legal regulations concerning fourth century A.D. gold mines show that the miners were encouraged, having paid eight scripuli of gold to the fisc, to sell more of the extracted metal to the State at an "appropriate price" (pretio proprio). (41) Since the State always needed a large and regular supply of raw metal, the sale of metals directly to the State may have been officially encouraged. (42)

Some recently discovered tin ingots, whose origin was possibly the Lusitanian tin mines of Albuquerque (E 06), help to throw some light on the organization of tin production during the reign of the emperor Claudius. (43) A large proportion of the ingots bear the stamp of L. VALERIVS AVG(ustae) L(ibertus) A COM(mentariis), a freedman of Claudius' first wife, Messalina. His title suggests that he was an official assistant to the provincial procurator. (44)

But most of the legible ingots are stamped with more than one name : for example, one is stamped in two places with the name L. VALERIVS AVG. L. A COM., in three places with L. CORNE(lius) VEG(etus) and in one place with L. AVR(elius); another has the names L. VAL(erius) twice, ANTVL(lius ? or Antuleius or Antullus) twice, Q. CARINAT(ius) once, L. AVR(elius) once, as well as M.T.S. significantly superimposed over the name of the imperial freedman. (45) This is significant because it suggests that the name of the imperial freedman was stamped first - possibly at the time of smelting. Various reconstructions are possible : one would be to envisage the other names as representing an association of people involved in the extraction and/or the smelting of the ore, who then handed this tin over to the fisc in payment of taxes. At this stage the imperial administrator has his stamp made on the ingots. The overstamping of the initials M.T.S. might then represent a third stage : namely, the selling-off of the tin by the provincial fisc to a private shipper, who was responsible for its transfer from the production area to a market possibly somewhere in southern Gaul, given the location of the wreck.

Alternatively, the mines may have been controlled by the imperial fisc, who took a proportion of the metal extracted in return for the right to work the mines. On this interpretation the first stamp on the ingot would have been that of the imperial administrator, who then sold off the metal to the other private individuals whose names were subsequently stamped on the ingots and who arranged for its sale on whichever private market they saw fit. It is impossible to prove decisively which of the two scenarios is more likely; but given that we are dealing with tin, as opposed to gold or even silver, I would be tempted to suggest that the first is more likely, viz. that the mines were privately owned during the reign of Claudius.

This mode of operation is perhaps confirmed by some lead ingots from the shipwreck discovered at Ses Salines, Majorca, which suggest that the lead mines in question were privately owned and that the mine owner had first smelted the metal and then sold it (or handed it over in payment of tax) to the provincial fisc. For the ingots were first stamped with the names of private individuals, presumably the mine owners (for example, P. AEMILI. GALLICI, N.MEVI. APRI, Q. CORNVTI, L. MANLI.), and then restamped with an imperial mark (VESP. AVG. or IMP. CAES.), once they had been sold to the State. (46)

The two sets of ingots suggest that even if the State did not own a particular mine, it was always a potential market for the metals produced. So much of the mineral extracted from the subsoil of Lusitania would have ended up in the hands of the provincial procurator. But there would also have been a market for metals in the various urban centres of the province for conversion into building materials, tools, as well as luxury items for the municipal elite, once the needs of the Roman State had been satisfied. The use of metal in bar or ingot form should not be underestimated as a means for the hoarding of wealth. Many coin hoards from the Iberian peninsula consist not only of coins, but sometimes coins associated with precious metal : either in the form of jewelry or of ingots. (47) A good example of the latter is provided by a hoard recently discovered at Castro de Alvarelhos, Santo Tirso, near Oporto (just to the north of the province of Lusitania), where nine silver ingots are associated with Roman <u>denarii</u> of Republican and Augustan date, the latest dated to c. 28 B.C. (48) Thus some silver was kept and put into local use; not all metal extracted was surrendered to the provincial fisc.

Copper ingots have also been discovered (whose origin was either the south of Lusitania or the Sierra Morena mines in Baetica) with official stamps bearing the name of a procurator. (49) Another (found on the beach at Marseillan, Hérault) is stamped with the name of a freedman procurator, Felix, and an imperial slave, Telesphorus. (50) These would all suggest that by the mid second century A.D. some of the copper mines of southern Iberia had passed into imperial hands or at least that the metals had been transferred to the Roman State after their extraction.

iv. The Chronology of Lusitanian Mines

Some attempt must be made to assess the chronological development of mining in the province. Literary evidence is on the whole unsatisfactory, although references to mining activity in Strabo suggest that operations had already begun by the first century B.C. To present a more precise and nuanced picture, it is necessary to turn once again to the archaeological material from individual sites. Again problems arise, since material finds have been sporadically and unsystematically recorded and later periods of exploitation have tended to obliterate or obfuscate evidence of earlier workings. Also the discovery of Roman material from a mining area might not necessarily prove that the mine in question was operational at that particular time, but might only reveal that the area in general was occupied.

The most convenient way of presenting the chronology of individual sites is in catalogue form (see Appendix I). A summary of this evidence is given in Table 3.6.

TABLE 3.6 DATABLE ROMAN MATERIAL FROM LUSITANIAN MINES

A. CONVENTUS PACENSIS

P 09	São Estevão (Cu)	Roman lamp & coins : C4 A.D.
P 25	Daroeira, Alcoutim (Cu)	Coin of Antoninus Pius
P 30	Brancanes (Cu)	Coins of Augustus, Vespasian and Probus
P 33	Aljustrel (Cu/Ag/Au?/Fe)	Roman pottery : C1 B.C., C1, 2, 3, 4, 5 A.D. Coins : late C3 A.D.

P 36	Serra da Caveira (Cu/Ag/Au)	Roman lamps : Augustan, C4 A.D. (uncertain if from mine itself)
P 39	Serra dos Monges (Cu/Fe)	Lamp : C1 A.D.

B. CONVENTUS EMERITENSIS

E 06	Tres Arroyos (Sn)	Pottery : C1/C2 A.D.
E 17	Algibe (Ag/Fe)	Coins of Trajan
E 19	Plasenzuela (Ag/Pb/Cu)	Lamps : 1st half C1 B.C. Inscription : C1 A.D.
E 22	Nava de Ricomalillo (Au)	Inscriptions : C1, C2 A.D.

C. CONVENTUS SCALLABITANUS

S 29	Rio Maçainhas, Belmonte (Sn)	Coins : 1st half C1 A.D.
S 41	Serra da Lousã (Au/Ag)	Lamps : 1st half C1 A.D., C1 A.D., C2 A.D. Inscription : C1/C2 A.D.
S 38	Mina do Braçal (Pb)	Lamps : C1, C2 A.D.

That only such a small proportion of Lusitanian mines has provided any datable material is eloquent testimony to the extremely partial nature of the archaeological record. But the presence of datable material from the early period of Roman occupation of the province (for example, the lamp dated to the first half of the first century B.C. from Plasenzuela (E 19)) suggests that exploitation of the mines began before the official organization of the province under Augustus. Furthermore, at Vipasca (P 33) excavations have revealed that after some activity in the late Bronze Age the site at Algares was only re-occupied in the second century B.C. and substantial mining activity seems only to have begun again in the latter part of the first century B.C. (51) This prompt Roman interest in exploiting the mineral resources is also suggested by the adoption of many native Iberian words into the technological Latin vocabulary of Roman mining, as used not only in the elder Pliny's account of Spanish mining (for example, _balux_, _balluca_, _palae_, _palacurnae_, _psalacurnae_, _gangadia_, _arrugiae_), but also in the legal texts from Vipasca itself - for example, _rutramina_, _ubertumbus_, _lausiae_, _pittacaria_, _echolae_, _ternagus_. (52)

Various hoards of Roman Republican coins have also been discovered in mining zones. At this stage Roman coins are most likely to have been hoarded by Roman/Italian settlers (mainly consisting of army veterans who decided not to return to Italy after their period of service in the Iberian peninsula), rather than by locals. (53) Thus the hoards illustrate the early involvement of Romans in the mines before the official organization of the province. Again the evidence is best tabulated (Table 3.7).

TABLE 3.7 ROMAN REPUBLICAN COIN HOARDS FOUND IN MINING AREAS OF LUSITANIA

LOCATION OF HOARD/ ROMAN MINE	REF.	DATE (latest coin)	REF. for DATE
1. Penhagarcia S 15-22	RRCH 191 Raddatz, 282	104 B.C.	RRC no. 317
2. Idanha-a-Velha (1) S 15-22	Hipolito 96	c. 92 B.C.	FC III 199
3. Idanha-a-Velha (2) S 15-22	Numisma, 1980, 103-118	100 B.C.	RRC no. 330/1a
4. Idanha-a-Nova S 15-22	RRCH 229	89 B.C.	RRC no. 344
5. Atalaia, Monforte S 15-22	Hipolito 97 Raddatz, 278	?	----
6. Monforte da Beira S 15-22	Hipolito 98 Raddatz, 278-80	?	----
7. Monte do Sodes P 23	ARA II 387	?	----
8. La Presa, Meimoa S 26	XIV Congr.nacional de Arq., 1975, 1201-3	?	----

9. Quinta da Madeira Covilhã S 30	*Coin Hoards*, 7, 1985, No. 230	14 B.C.	*RIC* 327

Outside Lusitania

1. Sierra Morena (1)	*RRCH* 186	106 B.C.	*RRC* no. 311
2. Sierra Morena (2)	*RRCH* 196	102 B.C.	*RRC* no. 322
3. Rio Tinto	*RRCH* 194	102 B.C.	*RRC* no. 321
4. El Centenillo	*RRCH* 181	110-9 B.C.	*RRC* no. 300

NOTES TO ABBREVIATIONS

ARA = M.L. Estacio da Veiga Affonso dos Santos, *Arqueologia romana do Algarve* (2 vols.), Lisbon, 1971, 1972

RRCH = M.H. Crawford, *Roman Republican Coin Hoards*, London, 1969

RRC = M.H. Crawford, *Roman Republican Coinage*, Cambridge, 1974

Hipolito = M. de Castro Hipolito, 'Dos tesouros de moedas romanas encontradas em Portugal', *Conimbriga*, 2-3, 1960-61, 1-166

Raddatz = K. Raddatz, *Die Schatzfunde der iberischen Halbinsel*, Berlin, 1969

Hoards 1, 2, 3, 4, 5 and 6 are all from a zone of intense gold extraction from the Erges, Aravil and Ponsul tributaries of the Tagus, where eight Roman mining sites have to date been discovered (S 15-22). Hoards 1, 5 and 6 were found together with collections of gold and silver jewelry and ornaments, probably wrought from locally extracted metal. These hoards can all be dated to the very end of the second century B.C. or to the first quarter of the first century B.C. Hoard 7 was found near the copper mines of Cova dos Mouros, Alcoutim (P 23). Crawford has recently argued that these coin hoards might reflect a possible closure of the mines in the light of slave revolts. (54) To my knowledge there is no firm evidence whatsoever for slave revolts in the area, and it begs the question as to whether slave labour was used in the mining operations at this early date. A slightly more plausible hypothesis might be trouble in the area during the Sertorian War of the late 80s. Domergue has argued that this conflict caused a break in production in some of the mines of the Sierra Morena to the south of Lusitania (55). But on the whole the evidence is too thin to allow these rather fanciful hypotheses to be accepted.

This prompt exploitation of the mineral resources by Roman/Italian settlers is consistent with evidence from other parts of the peninsula. The silver mines at Carthago Nova were being exploited by Italians soon after the expulsion of the Carthaginians. (56) In the north-west the Romans took an immediate interest in the extraction of gold, once the area had been subdued, as was the case also in Britain. (57)

The Romans do not seem to have initiated the mining, but rather to have developed and extended the mining and metallurgy which had already formed an important part of the local economy in the late Bronze and Iron Ages. The thriving metallurgy in these periods is strikingly illustrated by the widespread finds of gold and silver plate and jewelry throughout the province. (58) In addition, Hannibal is said to have sunk mining shafts in the Iberian peninsula and slag has been discovered from pre-Roman workings at Rio Tinto. (59) As for Lusitania, inscriptions in late Phoenician script and a Phoenician amphora are reported from Vipasca (P 33) and late Bronze Age areas of mining activity have been located. (60) The necropolis discovered at Ourique, only some 40 km. to the south of Vipasca, has provided clear evidence of Phoenico-Punic interest in the area, with its inscribed stelae and Phoenico-Punic ceramics. (61) Further sites were first exploited in the pre-Roman period : for example, Arregata, Aljezur (P 04), Cisterna dos Cães (P 08), São Estevão (P 09), Vale de Pegas (P 12), Vendinha do Esteval (P 15), Serra da Caveira (P 36) and Serra dos Monges (P 39).

The extensive excavation and field survey carried out recently in the Sierra Morena mining zone reveals that in the Republican period the scale and general pattern of mining was not fundamentally altered by the arrival of the Romans; it was only after Augustus that new mines were opened and many small mines were expanded into larger-scale workings. (62) The need for peaceful conditions was just as important for mining as for other economic activities. The campaigns against Sertorius caused the foundries attached to the lead mines at Azuaga, Badajoz to be relocated on a fortifiable hill-top site until 76 B.C., when they reverted to lower ground closer to the mines themselves. (63) Similarly, the campaigns of the Civil War between Caesar and Pompeius caused a break in production in some mines of the Baetis valley. (64)

As for Lusitania the nature of the evidence does not allow such a detailed and nuanced picture to be drawn. The mines of Vipasca (P 33), however, prove an exception. The main dating evidence comes not from the mines, but from the cemetery of the mining community at Valdoca. The goods from the 496 graves excavated suggest that it was in use from the start of the first century A.D. until the second half of the third century. (65) This evidence is not sufficient to date the overall operation of the mines, but the general pattern suggested by this evidence is confirmed from elsewhere on the site. Material from later than the mid third century is rare, although some plates of the late fourth/early fifth century A.D. have been found in the recently excavated "House of the Procurator". Recently excavations have been carried out in part of the mining settlement, where modest dwellings have been revealed, arranged in something approaching the traditional urban grid-pattern of <u>insulae</u>. The first phase of occupation seems to have consisted of wooden buildings dated to the first century B.C., which were later superceded by stone buildings. The majority of finds can be dated to the second and third centuries A.D., but material exists from phases stretching from the first century B.C. to the fourth A.D. (66) The general conclusion seems to be that the mine was at its peak in the first two centuries A.D., it underwent a decline in production in the third century, but continued to be exploited well into the later Empire.

Further confirmation of a decline, and subsequent recovery, is provided by an inscription from Vipasca, that can be dated to A.D. 146, 173 or 235, depending on which consular name is restored in the damaged last line. The inscription commemorates the erection of a statue by some "coloni Metalli Vipascensis" to an imperial freedman procurator of the mines, styled "restitutor metallorum". (67) The arguments of Wickert, followed by the editors of <u>PIR</u>, in favour of a date of A.D. 235 (on the rather subjective grounds of letter-forms) are not conclusive. Moorish raids into the south of Lusitania and Baetica are attested for the 170s. (68) If these raids affected

Vipasca, the title "restitutor metallorum" would not be inappropriate for A.D. 173. After this restitution mining did continue, as the grave goods from Valdoca and other material from the mines suggest, but on a reduced scale.

Most other mines with datable material can be shown to have been operational in the first and second centuries A.D. It is thus probable that this was the most flourishing period of Lusitanian mining. But material from the later Empire is forthcoming, which suggests that not all the mines were in decline. The military site at Las Merchanas, Lumbrales (E 33) was certainly occupied in the mid third century and since its function was to supervise the tin mining of the area to the west of modern Salamanca, we must, therefore, suppose that tin was still being mined in this period. (69) This reduction in the scale of Lusitanian mining in the later Roman Empire may be paralleled from elsewhere in the Iberian peninsula : for example, at Rio Tinto and in the north-west gold mines. (70)

References to late-fourth century gold mining in the Theodosian Code all concern merely the eastern part of the Empire : Thrace, Macedonia, Illyricum, inland Dacia, Upper Moesia and Dardania, Pontus and Asia. (71) This has led to the assumption that mining in the western Empire was in a state of decline and decay. The arguments of Blazquez in favour of a renewal of Iberian mining in the fourth century A.D. are not entirely convincing. (72) As already stated, it is dangerous to use the evidence of Roman poets too indiscriminately. References to "aurifer Tagus" in such works can never prove that alluvial gold production still took place at this late date, since the idea had clearly become part of the literary tradition. (73) The other major strand in his argument - namely that the Roman administration expended much effort on keeping the roads leading to the north-west gold mines in a good state of repair - is plausible, but constitutes insufficient grounds for accepting his arguments : there were other reasons for keeping the roads open, especially since the only legionary base in the peninsula was located in the north-west, at León.

However, there is scattered archaeological evidence for activity in Roman mining areas in the fourth century, as we have seen for Vipasca. Similarly evidence, no matter how scant, is forthcoming from other mines of the peninsula. It is thus unwise to think of total decay. Mining did continue right up to the end of the Roman period of control of the peninsula, but under a different mode of exploitation from that prevalent in the early Empire. The operation of large-scale mining districts had for various reasons become impractical in a changed economic world. Mining did not cease, the demand for metals of all sorts being just as pressing. But there was now more emphasis on smaller units of exploitation, less directly controlled by the State. (74)

NOTES : CHAPTER THREE

* References to individual mining sites in Lusitania are not given in these notes. The reader is referred to Appendix I, where bibliography is given under each site. VIP A refers to the Lex Metalli Vipascensis, VIP B to the Lex Metallis Dicta (both of which are translated in Appendix II).

1. See J.F. Healy, Mining and Metallurgy in the Greek and Roman World, London, 1978, esp. chapter x.

2. Dio 52.28.4

3. So, for example, M.P. Charlesworth, Trade Routes and Commerce of the Roman Empire, Hildesheim, 1961, 157; J. Harmand, L'Occident romain, Paris, 1960, 382.

4. 3.2.8 (146)

5. Silver: 5.35-38; tin: 5.38.4

6. P.R. Lewis & G.D.B. Jones, 'Roman gold-mining in north-west Spain,' JRS, 60, 1970, 169-185; R.F.J. Jones & D.G. Bird, 'Roman gold-mining in north-west Spain II', JRS, 62, 1972, 59-74; C. Domergue, 'Introduction à l'étude des mines d'or dans le nord-ouest de la péninsule ibérique' in Legio VII Gemina, León, 1970, 255-286; C. Domergue & G. Hérail, Mines d'or romaines d'Espagne : le district de la Valduerna (Leon) : étude géomorphologique et archéologique, Toulouse, 1978; C. Domergue & P. Sillieres, Minas de oro romanas de la provincia de León I. La Corona de Quintanilla : excavaciones 1971-1973; Las Coronas de Filliel, Boisan, Luyego I & II : exploraciones 1973 (= EAE, 93), Madrid, 1977; C. Domergue & T. Martin, Minas de oro de la provincia de León II (= EAE, 94), Madrid, 1977.

7. G.D.B. Jones, 'The Roman mines at Rio Tinto', JRS, 70, 1980, 146-165; A. Blanco Freijeiro, 'Antiguedades de Rio Tinto', Zephyrus, 13, 1962, 31-45; J.M. Luzon

Nogue & D. Ruiz Mata, 'El poblado minero de Rio Tinto', Habis, 1, 1970, 125-138; B. Rothenberg & A. Blanco Freijeiro, Studies in Ancient Mining and Metallurgy in South-West Spain : explorations and excavations in the province of Huelva (= Metal in History, 1), London, 1981.

8. T.A. Rickard, 'The mining of the Romans in Spain', JRS, 18, 1928, 129-143; J. M. Blazquez Martinez, 'Explotaciones mineras en Hispania durante la Republica y el alto Imperio romano : problemas economicos, sociales y tecnicos', AHES, 2, 1969, 9-68; id. 'Fuentes literarias griegas y romanas referentes a las explotaciones mineras de la Hispania romana' in La Mineria Hispana e Iberoamericana, León, 1970, I, 117-150; id. Historia economica de la Hispania romana, Madrid, 1978, 21-42 (Republic), 85-98 (first century A.D.), 144-56 (second century A.D.), 242-47 (late Empire).

9. 3.3.4 (153)

10. 3.3.5 (154)

11. 3.2.3 (142)

12. 3.2.9 (147)

13. NH 33.21.66

14. NH 33.21.78

15. R.P. Duncan-Jones, 'The wealth of Gaul', Chiron, 11, 1981 217-220, esp. 217, note 6 for the figures.

16. NH 34.47.156-157

17. NH 34.49.164

18. NH 4.22.118

19. F.J. Fernandez Nieto, 'Aurifer Tagus', *Zephyrus*, 21-22, 1970-1971, 245-259, esp. 245-7.

20. contra J.M. Blazquez, art.cit.2 (n. 8), 148 and C.R. Whittaker, 'Inflation and the economy in the fourth century A.D.' in C.E. King (ed.), *Imperial Revenue, Expenditure and Monetary Policy in the Fourth Century A.D.* (= BAR Int. Ser., 76), Oxford, 1980, 1-22, esp. 5.

21. Aljustrel : VIP A, 7. i, lines 46-50; Laurion : R.J Hopper, 'The Laurion mines : a reconsideration', *ABSA*, 63, 1968, 293-326, esp. 309.

22. 3.2.8 (146)

23. J.F. Healy, op.cit. (n.1), 179-180; G.C. Boon, 'Aperçu sur la production des métaux non ferreux dans la Bretagne romaine', *Apulum*, 9, 1971, 453-503, esp. 465-467.

24. C. Domergue, 'Les Planii et leur activité industrielle en Espagne sous la république', *MCV*, 1, 1965, 9-27 & 'Les lingots de plomb romains du Musée archéologique de Carthagene et du Musée Naval de Madrid', *AEA*, 39, 1966, 41-72; id., *Les lingots de plomb romains estampillés d'Espagne* (forthcoming).

25. For ingots : J.C. Allan, *Considerations on the Antiquity of Mining in the Iberian Peninsula* (= Royal Anthropological Institute, Occasional Paper 27), London, 1970, 16; D.F. de Almeida, 'Mineração romana em Portugal' in *La Mineria Hispana e Iberoamericana*, Leon, 1970, I, 195-220, esp. 213; A. Viana, O. da Veiga Ferreira & R.F. d'Andrade, 'Exploração das minas de Aljustrel pelos romanos', *Arquivo de Beja*, 13, 1956, 3-20, esp. 15. For metals extracted at Vipasca see now C. Domergue, *La mine antique d'Aljustrel (Portugal) et les tables de bronze de Vipasca* (= Publication du Centre Pierre Paris, 9), Paris, 1983, 24-30.

26. G.D.B. Jones, art.cit. (n.7), 146-154 and Appendix B (p. 165); M.P. Jones in B. Rothenberg & A. Blanco

Freijeiro, op.cit. (n.7), 33-34.

27. For the mineral resources of modern Portugal see Direcção geral de minas e servicos geologicos, Carta mineira de Portugal (1:500,000), Lisbon, 1960.

28. D. F. de Almeida, 'Minas de ouro na Gallaecia portuguesa' in Legio VII Gemina, León, 1970, 287-301.

29. W. Borlase, Tin Mining in Spain, past and present (= Ancient Science Tracts 429), London, 1897; O. Davies, Roman Mines in Europe, Oxford, 1935, 103.

30. 3,000,000 tonnes of slag survive from Vipasca : C. Domergue, op. cit. (n. 25), 28; 750,000 tonnes from Sao Domingos : J. de Alarcão, Portugal romano (3rd ed.), Lisbon, 1983, 134.

31. See Chapter Two, pp. 8-11.

32. Slag amounts to 300,000 tonnes : J. de Alarcão, op. cit. (n. 30), 134; for the shafts see J.L. de Vasconcelos, 'Excursão arqueologica a Extremadura Transtagana', AP, 19, 1914, 300-323, esp. 310-311.

33. So A.H.M. Jones, The Roman Economy (ed. P.A. Brunt), Oxford, 1974, 68.

34. F.J. Sanchez Palencia, 'La explotación del oro en la Hispania romana : sus inicios y precedentes' in I Congreso Internacional de Mineria antigua en el mundo mediterraneo (Madrid, oct. 1985) (forthcoming). I am very grateful to Dr. Sanchez Palencia for letting me consult this article prior to publication.

35. Strabo 3.2.10 (148); J.J. van Nostrand, 'Two notes on the mines of Roman Spain', Pacific History Review, 4, 1935, 281-289, esp. 282.

36. For example see ILS 6870 for regulations concerning the Saltus Burunitanus. On parallels between the Vipasca texts and the regulations for North African

imperial estates see D. Flach, 'Die Bergwerksordnungen von Vipasca', Chiron, 9, 1979, 399-448, esp. 440-443.

37. J.A. Crook, Law and Life at Rome, London, 1967, 161.

38. Dig. 7.1.13.5

39. Tac. Ann. 6.19

40. VIP B, 2, line 12

41. C.Th. 10.19.3; CJ 11.7.1

42. J.A. Crook, op.cit. (n. 37), 162.

43. D. Colls, C. Domergue, F. Laubenheimer, B. Liou, 'Les lingots d'étain de l'épave Port-Vendres II', Gallia, 33, 1975, 61-94; D. Colls, R. Étienne, R. Lequément, B. Liou, F. Mayet, L'épave Port-Vendres II et le commerce de la Bétique à l'époque de Claude (= Archaeonautica, 1), Paris, 1977, 11-18.

44. The office of a commentariis can be paralleled from a mining region of Dalmatia, where an a commentariensis aurariarum Dalmatorum is attested : CIL III 1997 = ILS 1595.

45. D. Colls, C. Domergue et al, art.cit. (n. 43), 70 (no. 9), 67-9 (no. 12).

46. C. Veny, 'Diecesiete lingotes de plomo de una nave romana de Ses Salines (Mallorca)', Ampurias, 31-32, 1969-1970, 191-219.

47. K. Raddatz, Die Schatzfund der iberischen Halbinsel von Ende des dritten bis zur Mitte des ersten Jahrhunderts vor Chr. Geb. : Untersuchungen zur hispanischen Toreutik (= Madrider Forschungen, 5), Berlin, 1969.

48. Coin Hoards, 6, 1981, 24, no. 89.

49. F. Benoit, 'Nouvelles épaves du Provence', Gallia, 20, 1962, 154 proposes an origin in Onuba (Huelva); M. Euzennat, 'Lingots espagnols retrouvés en mer', Etudes classiques (Publication de l'Université d'Aix-en-Provence), 3, 1968-1969, 83-98 suggests an origin in Ossonoba (Faro, Portugal), followed by F. Laubenheimer-Leenhardt & H. Gallet de Santerre, Recherches sur les lingots de cuivre et de plomb d'époque romaine dans les regions de Languedoc-Roussillon et de Provence-Corse (= Revue archéologique de Narbonnaise, Suppl. 3), Paris, 1973, 36-37 (with figure 14bis) & 76-82. The poor state of preservation of the inscription at the vital point will not allow for any certainty on this matter. The ingots can now be dated to the reign of Antoninus Pius (see M. Euzennat, art.cit).

50. F. Laubenheimer-Leenhardt & H. Gallet de Santerre, op.cit. (n. 49), 53-55 (with figs. 23 & 24) & 82-84.

51. C. Domergue, op.cit. (n. 25), 30-32; C. Domergue & R.F. d'Andrade, 'Sondages 1967 et 1969 à Aljustrel (Portugal) : note préliminaire', Conimbriga, 10, 1971, 99-116.

52. Pliny : NH 33.21.70-78, on which see J.F. Healy, 'Problems in mineralogy and metallurgy in Pliny the Elder's Natural History' in Tecnologia economica e societa nel mondo romano, Como, 1981, 163-201, esp. 184-186. Vipasca : VIP A & VIP B passim.

53. M.H. Crawford, Coinage and Money under the Roman Republic : Italy and the Mediterranean Economy, London, 1985, 100-2.

54. ibid., p. 102.

55. C. Domergue, 'Un temoignage sur l'industrie minière et métallurgique du plomb dans la région d'Azuaga (Badajoz) pendant la guerre de Sertorius', XI Congreso Nacional de Arqueologia, Mérida, 1968, Zaragoza, 1970, 608-626.

56. Diod. Sic. 5.36; C. Domergue, art.cit. 2 (n. 23); A.J.N. Wilson, Emigration from Italy in the Republican Age of Rome, Manchester, 1966, 26-27.

57. North-west Spain : R.F.J. Jones, 'The Roman military occupation of north-west Spain', JRS, 66, 1976, 44-66, esp. 48. Britain : S. Frere, Britannia : a history of Roman Britain, London, 1975, 284; here the Dolaucothi gold mines are a good example : see P.R. Lewis & G.D.B. Jones, 'The Dolaucothi gold mines I : the surface evidence', Ant. J., 49, 1969, 244-272, esp. 260.

58. On which see K. Raddatz, op.cit. (n. 47) & M. Almagro Gorbea, El Bronce final y el periodo orientalizante en Extremadura (= Bibliotheca Praehistorica Hispana, XIV), Madrid, 1977.

59. Hannibal : Pliny NH 33.21.96; Rio Tinto : J.S. Richardson, 'The Spanish mines and the development of provincial taxation in the second century B.C.', JRS, 66, 1976, 139-152, esp. 140 & P. Saenz Fernandez, 'Metalurgia y comercio punicos en Sierra Morena' in Homenaje al Prof. Dr. Hernandez Diaz, I, Seville, 1982, 105-115.

60. O. Davies, op.cit. (n. 29), 118 (inscriptions); C. Domergue, op.cit. (n. 25), 35 (amphora), 36 (Bronze Age arrowhead).

61. C. de Melo Beirão, M.M. Alves Dias & L. Coelho, 'Duas necropoles da Idade do Ferro no Baixo Alentejo : Ourique (noticia preliminar)', AP, iii, 4, 1970, 175-219; see now the dramatic votive deposit including Phoenico-Punic glass and representations of the Punic deity Tanit : C. de Melo Beirão et al., 'Deposito votivo da II idade do ferro de Garvão : noticia de primeira campanha de escavações, AP, iv, 3, 1985, 45-136.

62. B. Rothenberg & A. Blanco Freijeiro, op.cit. (n. 7), 173.

63. C. Domergue, art.cit. (n. 55), esp. 615.

64. C. Domergue, 'Rapports entre la zone minière de la Sierra Morena et la plaine agricole du Guadalquivir à l'époque romaine', MCV, 8, 1972, 614-622, esp. 620.

65. O. da Veiga Ferreira & R.F. d'Andrade, 'A necropole de Valdoca (Aljustrel)', Conimbriga, 5, 1966, 1-6; J. & A. de Alarcão, 'O espolio da necropole luso-romana de Valdoca (Aljustrel)', Conimbriga, 5, 1966, 7-104.

66. Mining settlement : R. Parreira, 'Intervenção de emergencia no povoado mineiro romano de Vipasca - Aljustrel, distrito de Beja', Informação arq., 4, 1981, 135-44; id., 'O salvamento arqueologico de Vipasca : nota preliminar sobre a campanha de escavacoes de 1981', Arq. de Beja, iii, 1, 1984, 83-106. House of the Procurator : C. Domergue, op.cit. (n. 25), 36.

67. IRCP 121 = L. Wickert, 'Bericht uber eine zweite Reise zur Vőrbereitung von CIL II Suppl. 2', Sitz. der Preuss. Ak. der Wissen., 1931, 829-844, esp. 835-839. See also E. Cuq, 'Notes d'épigraphie et de papyrologie juridiques', NRHDFE, 1908, 285-312, esp. 306-310 & S. Lambrino, 'Catalogue des inscriptions latines du Musée Leite de Vasconcelos', AP, iii, 1, 1967, 123-217, esp. 130-131. Possible dates : A.D. 146 (Cn. Claudius Severus Arabianus : PIR 1027), A.D. 173 (Cn. Claudius Severus : PIR 1024) or A.D. 235 (Cn. Claudius Severus : PIR 1025).

68. SHA, M.Ant. 21.1; 22.11. Note also ILS 1327, 1354, 1354a. See J. Arcé, 'Inestibilidad politica en Hispania durante el siglo II d.C.', AEA, 54, 1981, 101-115, esp. 110-1 arguing for internal disturbances rather than Moorish incursions.

69. J. Maluquer de Motes, 'Excavaciones arqueologicas en el castro de "Las Merchanas" (Lumbrales, Salamanca)', Pyrenae, 4, 1968, 101-128, esp. 119; J.M. Blazquez, op.cit. (n. 8), 245-6.

70. Rio Tinto : G.D.B. Jones, art.cit. (n. 7), 161-163; for north-west Spain : C. Domergue, art. cit. (n. 6), 279.

71. CT 1.32.5 (dated to 29 July, A.D. 386).

72. op.cit. (n. 8), 243-6.

73. v. supra, p. 26 & note 20.

74. I shall discuss this fully in my forthcoming article, 'Mining in the later Roman Empire : continuity or disruption ?'.

CHAPTER FOUR

MINING

i. Economy and Society of Mining Communities

The province of Lusitania is fortunate to possess the two legal documents which were discovered in 1876 and 1906 in the slag heaps of Roman workings at Vipasca. In addition to the light they throw on legal and administrative matters, they are as revealing of the economic substructures and social makeup of a mining community. (1) For the purposes of this chapter the more localised set of rules contained in the Lex Metalli Vipascensis (FIRA I 105, henceforth VIP A, translated in full in Appendix II) are more relevant than the more widely applicable conditions of the Lex Metallis Dicta (FIRA I 104, henceforth VIP B, also translated in Appendix II).

In the Roman world large mining communities were administered independently from the nearest urban centre and did not form part of the territory of that centre. (2) The mines of Vipasca were not, therefore, part of the territory of Pax Iulia (Beja), but independent imperially-owned land. Thus they were not administered, as were other urban nuclei, by duovirs and a town council of decurions, but were under the direct control of a procurator metallorum, who was in turn subordinate, and accountable, to the procurator provinciae Lusitaniae et Vettoniae, resident in Augusta Emerita, the provincial capital. In the legal texts the community of Vipasca is referred to as a vicus, if the conventional restoration of the text is accepted, which had its own territorium or fines. (3) The vicus was equipped with a forum, where official documents and notices were displayed and statues of prominent figures were set up : for example, the statue erected in honour of the procuratorial "restitutor metallorum". (4)

A mining community such as Vipasca operated outside the normal economic system of the province. Thus the procurator at Vipasca let out contracts for the sole right to run various trading concessions : for example, shoemaking, barbering, fulling, auctioneering, Mining communities did not use standard Roman currency, but tesserae only valid within their territory. (5) This general exclusion from the wider monetized economy can be paralleled in the company shops that operated in later Cornish tin mines and in some Yorkshire colleries. (6)

Auctions were held not only for the letting of mining concessions, but also for the sale of slaves, mules, asses and horses. The right to conduct these auctions (the scriptura praeconii) was also leased out to a concessionary, as was the right to run the community's baths. (7) The nature of the contract for the latter is in many ways similar to the pottery manufacture leases from Oxyrhynchus in Roman Egypt. (8) In both cases the lessee leases the respective buildings and the associated equipment. At Oxyrhynchus the landowner also provided the raw materials : the clays, firing materials for the kilns, water for the cistern and pitch for covering the jars. (9) At Vipasca there is some doubt as to whether the lessee had to provide fuel for himself. It is stated that the lessee should always have an adequate supply of wood and that he should not be allowed to sell it to other members of the community, with the exception of the branch trimmings that were unsuitable for fuel. (10) As will be discussed in more detail below, the need for wood was of paramount importance in a mining community. It was used for building shafts, props and galleries in the mines themselves and, most essentially, for fuel to heat the smelting and processing furnaces. Much attention must have been paid to its supply. On the analogy of the evidence presented elsewhere in the legal texts, the procurator may well have let out another concession for the provision of the community's timber requirements. (11)

Furthermore, there is evidence that other Roman towns owned public woodlands in their territory, the wood gathered from which was used for public buildings and for fuelling the public baths. (12) An inscription from Aurgi (Jaén, Hispania Tarraconensis) records the donation of public baths by a local benefactor, who also provided roughly thirty-seven hectares of woodland to provide fuel for the baths, as well as an aqueduct to ensure the water supply. (13) This close connexion between running a bath complex and fuelling it would suggest that the two processes may not have been administered separately at Vipasca.

The presence of schoolmasters at Vipasca (as elsewhere in the Roman Empire exempt from taxation) emphasises that the community needed all the services and facilities of a normal urban centre. (14) In short, it required the same social and economic infrastructures.

The artefacts uncovered in the excavation of the cemetery of Valdoca give some impression of the range of goods that had to be produced in, or imported into, a mining community. The surviving pottery (terra sigillata tableware, thin-walled ware, coarseware, terracotta lamps and amphorae) and glass represents merely the tip of an iceberg. (15) Mining communities were in need of constant supplies of imported foodstuffs. A substantial proportion of manpower must have been involved with the provision, or production, of these goods.

Patterson has made some speculative calculations on manpower needs for ancient silver mining. The production of one tonne of silver required 500 - 1,000 man-work-years. (16) This large labour requirement was needed not only for mining, processing and smelting of ore, but also to provide the wide range of ancillary services that were necessary to keep such a work-force supplied with essential commodities and services. Polybius (as reported by Strabo) reckoned that the silver mines of New Carthage had a labour force of 40,000 men. (17) It is impossible that all 40,000 were involved in the actual mining of ore, since ancient mining shafts and galleries did not allow for work units of more than fifty men. (18)

A roughly comparable size of labour force was employed in sixteenth century mines at Potosi in Peru (now modern Bolivia). Of the 30,000 involved only 11,000 were engaged in the extraction and processing of ore (4,600 in the mines, 6,000 in washing and processing); the remaining 19,000 provided ancillary services for the miners, essentially being involved in agriculture or trade. Similarly at the mines of Zacatecas in Mexico only 1,700 (out of a total work-force of 5,000) were active in mining and processing. (19) Thus, at Potosi 63 per cent. and at Zacatecas 66 per cent. of manpower was devoted to ancillary services. The exploitation of mineral resources on a large scale in an area, therefore, greatly stimulated other economic activities - essentially trade and agriculture. (20)

Mines were often located in remote, mountainous regions, where agricultural conditions were unfavourable. (21) Much attention had then to be paid to the importation of the necessary supplies for the mining community. (22) Moreover, it would have been exceptional for the hinterland to have been capable of supplying sufficient manpower, given the often remote location of mines. A significant degree of immigration - whether forced or voluntary or both - to mining areas would, therefore, have been necessary. (23)

One source of involuntary labour was provided by criminals sentenced to work in the mines. Criminals were transported long distances from "non-mining" to "mining" provinces. (24) Also in the wake of the conquest and annexation of new territory the Romans transported some natives to work in mines. The toponym "Castellum Dalmatas" in the Remesiana district of Moesia Superior would suggest that certain Dalmatians were transported to this mining zone to work in the mines. (25) A similar process may be postulated for Lusitania. Strabo mentions that the Romans transplanted certain Lusitanians to the territory between the Tagus and the Anas, which he notes elsewhere as containing metalliferous mountains. (26) The recently identified series of forts between these two rivers may well reflect the need to keep a potentially hostile workforce under control. (27)

Voluntary immigration is well attested for the New Carthage silver mines, where individual Italians can be shown to have operated from the second century B.C. The names stamped on the lead ingots from these mines advertize the Italian origin of the exploiters. (28) Under the Empire funerary inscriptions from other mining sites also reveal a significant immigrant element. The inscriptions considered here are solely those which record the individual's ethnic (usually in adjectival form) : e.g. L. Cornelius Hispanus Cluniensis. No attempt is made to enter the dangerous jungle of Roman onomastics. As always with epigraphic evidence, the attested examples represent only those more privileged immigrants who could afford (or whose heirs could afford) to erect such a memorial. We may, therefore, be dealing only with the upper echelons of mining societies, investors in mining leases or successful traders, rather than with mine labourers.

The largest body of evidence comes not from Lusitania, but from the Sierra Morena mining district of Baetica. The evidence is best tabulated (Table 4.1).

TABLE 4.1 IMMIGRATION TO MINING COMMUNITIES

A. Immigrants in Lusitanian Mines

MINE	ORIGIN	NAME	REF.
1. Plasenzuela (E 19)	Turgalium	...Libaeco Caenici f.	HAEp 2674 AE 1977, 420
2. Nava de Ricomalillo (E 22)	Clunia	L.Cornelius Hispanus	BRAH 1903, 536
3. " " "	Avila	Maeso Burri f.	BRAH 1900, 175-6
4. " " "	Toletum	Anius Alpeti f.	BRAH 1897, 426-8
5. Villa del Rey (E 07)	Institienses	Accius Bouti f.	Zephyrus, 1970-71, 329-347

B. Lusitanian Emigrants in Iberian Mines

1. Rio Tinto	Emerita	L.Helvius Lupus	O. Davies, RME, 128, n. 1 (b); Zephyrus, 1962, 42 & fig. 12
2. Rio Tinto	Emerita	Probus	ibid.
3. " "	Emerita	Helvia Secundilla	ibid.

4.	"	"	Talabriga	Fuscus Fucini f.	O. Davies, RME, 128, n. 1 (c)
5.	"	"	(T)alabriga	Sutrio Calli f.	ibid. 128, n. 1 (d)
6.	"	"	Olisipo	?	CIL II 959
7.	"	"	Olisipo	L.Iulius Reburrinus	Habis, 1970, 137-8
8.	"	"	?Lusitania/ ?Galicia	Camalus	Blazquez, La Mineria I, 140
9.	Mons Marianus		Conimbriga	Vegetus Aviti f.	EE IX 32
10.	Santa Marina de Somoza, Teleno, Galicia		Lusitania	?Lusegicar...Meio ..nomiae F..nginus	HAEp 2145
11.	Plasenzuela (E 19)		Turgalium	...Libaeco Caenici f.	HAEp 2674

Of all Iberian mines Rio Tinto has the largest body of evidence of all Iberian mines for immigration. Immigrants are attested from Lusitania (Table 4.1, B, 1-8), but also from Asturia, (29) Arabriga (30) and Castellum Berense (31) in Hispania Tarraconensis, from Serpa in Baetica (an important mining area itself) (32) and from Turobriga, the location of which is uncertain. (33)

Whether all these immigrants moved to Rio Tinto to become involved in mining itself or in ancillary services for the mining community is, for the most part, uncertain. But the recent discovery of a second immigrant from Olisipo throws light on possible reasons for immigration. (34) L. Iulius Reburrinus came to Rio Tinto not to exploit the mineral resources, but to produce or sell mining lamps, twenty-six of which have been discovered with the stamp L.I.R. (35) This firm of lamp manufacturers supplied other mines, since examples have so far been discovered at the mines of Cerro Muriano (Corduba), Poderosa (Campofrio) and Aljustrel (P 33). (36) The distribution of his products raises the question as to whether Reburrinus was an itinerant potter or part of a firm of potters with branch workshops. (36) Whichever is the case, this is a clear example of increased economic sophistication being caused by an intensification of mining after the creation of the Roman province.

The evidence for immigration to Lusitanian mines is less striking, but the need for additional manpower would have been just as pressing. The evidence presented in Table 4.1 also suggests that it was the agriculturally unfavourable areas of the Iberian peninsula that supplied the greater proportion of immigrant manpower. It also suggests that it was native, freeborn individuals who decided to migrate to mining zones : for example, such obviously free, indigenous characters as Maeso Burri f. and Anius Alpeti f. from Nava de Ricomalillo. Sometimes an indigenous name in Romanized form occurs : for example, L. Aiosecus from the same mine or C. Vibius Paternus, who dedicated an altar to an indigenous deity at the Serra da Lousa gold mines (S 41). (38) Men of freeborn status are by far the largest group represented epigraphically at Vipasca (see Table 4.5).

After the provincial capital, Augusta Emerita, the urban centre of the Igaeditani (modern Idanha-a-Velha, ancient name unknown) can boast the next largest number of immigrants represented in the inscriptions of any Lusitanian town (see Table 4.2). This may reflect mere chance survival of evidence : Idanha-a-Velha failed to develop in the Middle Ages as a significant urban centre, even though it had been the seat of a Visigothic bishopric, and so did not need to reuse Roman inscriptions for later building. (39) But Idanha was located right in the centre of a zone of intense Roman gold workings along the Tagus and its tributaries. On the basis of archaeological finds it seems that the whole region was more densely settled in the Roman era than in the Iron Age. (40) Furthermore, Idanha attracted immigrants not only from other regions of the province of Lusitania, but also from further afield, notably Clunia in Hispania Tarraconensis. Of the Lusitanian immigrants most arrived from towns or civitates in the north of the province, their names publicizing their Celtic origins.

The most attractive explanation for this attraction of manpower would be the widespread mining activity around Idanha. An illustration of the scale of production in this region is provided by another inscription which records the dedication of a thanks-offering to Jupiter following the discovery of 120 pounds of gold in the Serra do Monsanto. (41) This amount is roughly equivalent to 540,000 sesterces, enough to build a temple and a theatre on the basis of costs from the African provinces under the Principate. (42) If they were attracted to become involved in this mining, this is a good illustration of one effect of the intensification of mining on a provincial society : it encouraged a good deal of geographical, and possibly social, mobility.

TABLE 4.2 IMMIGRANTS AT IDANHA-A-VELHA

	ORIGIN	NAME	REF.
A. HISPANIA TARRACONENSIS			
1.	Clunia	C. Fabius C.f. Vernus	Egitania, no. 81
2.	Clunia	Tuatro Frontonis f.	CIL II 450
3.	Clunia	Aternus A...cae f.	AE 1967, 148
4.	Clunia	L. Cornelius Q.f.	AE 1967, 164
5.	Libia	Arrenus Crescentis f.	CIL II 439
B. BAETICA			
6.	Italica ?	L. Cocceius	Egitania, no. 69

C. LUSITANIA

7.	Emerita	C. Furius Eutyches	Egitania, no. 69
8.	Emerita	C. Furius Lycius	Egitania, no. 70; HAEp 1119
9.	Salmantica	L. Antius Avitus	CIL II 438
10.	Talabara	Maeilo Camali f.	CIL II 453
11.	Lancienses Oppidani	Albinus Tangini f.	AE 1967, 147
12.	Lancienses Oppidani	Annius Valens	Egitania, no. 29
13.	Lancienses	Aunia Arontoni Celtiaci f.	Egitania, no. 36
14.	Meidubriga	Hispanus Tangini f.	AE 1977, 362
15.	Meidubriga	Proculus Meidubri	CIL II 458

N.B. Egitania = D.F. de Almeida, Egitania : Arqueologia e Historia, Lisbon, 1956

Just as a number of people from Clunia seem to have left their birthplace to seek employment in mines elsewhere in the peninsula, a group of people from Bracara Augusta (modern Braga) moved south to settle in and around Pax Iulia, possibly to work in the mines at Vipasca. (43) For a "sodalicium Bracarorum", a social club for emigré Bracari, is attested at Pax Iulia (44), while a certain Ladronus Dovai Bracarus from Castellum Durbedense (in the vicinity of Bracara) died at a Roman villa site at Franciscos, Garvão, Ourique, only 40 km. from Vipasca. (45).

The increased extent and scale of mining in Lusitania under the Romans must have had a profound effect on the local economy. One major repercussion throughout the peninsula was the employment possibilities that the more systematic exploitation of mineral resources provided. As we have seen, the employment was not necessarily in mining alone, but also in a plethora of ancillary tasks and services needed to maintain a large mining community. The prospect of employment seems to have caused a significant degree of mobility of personnel both into and out of Lusitania. This mobility in turn must have helped to create a higher degree of homogeneity among peoples and ethnic groups in an area of the Roman Empire previously marked by its very ethnic and cultural diversity. Furthermore, increased mobility and increased contact with Roman administrators, soldiers and settlers in mining zones would have hastened the process of Romanization of the indigenous population.

The legal texts from Vipasca also provide good evidence for the social makeup of a mining community. The following are mentioned : the procurator metallorum, imperial slaves and freedmen working for the procurator, free tenants (coloni) and their associates (socii), slaves, children, slaves and hired labourers employed in processing ore and in the stone quarries, women and, finally, soldiers.

First, procurators. From the admittedly scant evidence for individual procurators of Vipasca, it would seem as if they were appointed in general from among imperial freedmen. Ulpius Aelianus, the procurator to whom the Lex Metallis Dicta is addressed, seems (on onomastic grounds) to have been a freedman of the emperor Hadrian. (46) The name of the procurator styled "restitutor metallorum" is not preserved on his statue base, but he was explicitly an imperial freedman. (47) A third, Saturninus, is known from a marble statue base from Pergamum. Habicht has convincingly argued that he was an imperial freedman from the nature of his other positions. (48)

This tendency for imperial freedmen to be appointed from elsewhere in the Iberian peninsula : so, for example, M. Ulpius Aug. lib. Eutyches, the procurator metalli Albocensium in Gallaecia, (49) T. Flavius lib. Polychrysus (50) and Dorotheus Aug. lib., (51) procurators of the Mons Marianus mines in Baetica, or Pudens Aug. lib., procurator at Rio Tinto. (52) Although the evidence is scant, it also seems that there might have been certain mining specialists : Saturninus, already mentioned as a procurator of the Vipasca mines, had previously held office as procurator metallorum in the gold mining district of north-west Spain; similarly, C. Iulius Silvanus held procuratorships in Gaul, in north-west Spain and in the mining district of Dalmatia. (53)

No mining procurator of equestrian rank is yet known from the Iberian peninsula, but it is known that in the Pannonian iron-mines, for example, there was a change from imperial freedmen to equestrians in the third century A.D. (54) The absence to date of equestrian procurators from the Iberian peninsula may merely reflect the generally scant survival of third century evidence. Alternatively, it might reflect a change in the way mining was organized. It seems that the special procuratorship of Asturia and Gallaecia (as distinct from the rest of Hispania Tarraconensis) disappeared at the start of the third century A.D. Domergue has seen this as a symptom of the decline in mining in the area. (55) But a section of the Theodosian Code suggests that in the later fourth century decurions were fulfilling the role of mining procurators. (56) In other words, there had been a change in the way in which mining was organized. Perhaps there were similar changes in the Iberian peninsula.

Secondly, the imperial slaves and freedmen working for the procurator were presumably technical assistants, such as *tabularii* or *commentarienses*. The procurator himself was only responsible for the financial and juridical administration of the mining district. Problems of a technical nature were dealt with by such *tabularii* or *commentarienses*. (57) Stamps bearing the name of an *a commentariis* been found on tin ingots from the Port-Vendres II wreck, while other ingots bear witness to imperial slaves. (58) This would all suggest that such officials had a role to play in the supervision of the smelting and processing of the metals.

Thirdly, children and soldiers, who at Vipasca gained free admission to the community's baths. Children were certainly employed in ancient mining. (59) The narrowness of the galleries and shafts meant that children were very much at a premium, as they were later in British coal mines. A funerary relief from Baños de la Encina (Cartagena) depicts a child holding mining tools. (60) Although the child concerned was too young to be involved himself, the suggestion of the relief is that children were involved, as soon as they were physically able. The fact that they were granted free bathing rights suggests that they played a valuable working role within the mining community at Vipasca.

The presence of soldiers is consistent with what is known from mining communities in other parts of the Roman world. Tacitus talks of legionaries being involved in the mining of silver in Germany in the reign of Claudius.

(61) The military organization of north-west Spain was in part geared to the supervision of the gold mining. (62) In Britain military forts were located at Pumpsaint near the Dolaucothi gold mine and near the Charterhouse lead mines and legionary stamps are to be found on lead pigs from the Mendip lead mines. (63) In Egypt a centurion of the Legio XV Apollinaris was in charge of the quarrying at Mons Claudianus. (64) Finally, in Pannonia Procopius refers to various castella, which were probably linked with Roman iron extraction. (65)

From Lusitania there is additional evidence for military links with mining. In the conventus Emeritensis the site at Las Merchanas (Lumbrales, Salamanca) housed a military garrison in the third century A.D. to control the tin mining of the region. (66) In the conventus Pacensis a series of seventeen fortresses has recently been identified. (67) Maia locates the most northerly example at Mangancha, just to the north of the Vipasca mines. However, his identification of this site as a fort seems dubious; rather it was a Late Bronze/Iron Age hill-fort, so common in the Celticized parts of the Iberian peninsula. He has located others with greater certainty in proximity to Roman mines : for example, the fort at Castelo de Almodovar is near the copper mines at Almodovar (P 31) and the two forts of Castelo de Alcaria Cova and Castelo de Guerreiros do Rio are near the cluster of copper mines around Alcoutim (P 21-29).

The route that these forts seem to be protecting would have been ideal for transporting metals from the mines of the Alentejo and north-eastern Algarve to the port of Myrtilis (Mértola) on the Guadiana. They seem to have been occupied from the first century B.C. to the second century A.D. The theory that they were built to check some politico-military troubles in the area (perhaps in the Sertorian period) is difficult to substantiate, since there is no evidence that there was any fighting in this area. (68) It would also be hard to explain the homogeneity of the group on the grounds that they were all constructed to deal with isolated disturbances. Their connexion with the transport of minerals from the mines of the interior to the river port of Myrtilis seems a more plausible hypothesis, bearing in mind the close connexion between military forts and mines in other parts of the Roman Empire.

Other known military forts in the province were also established in areas of mining : the forts of Castra Servilia and Castra Caecilia, contributi to the colony of Norba Caesarina (Cáceres) according to Pliny, could have supervised mining at Plasenzuela (E 19), which was operational by the first half of the first century B.C. (69) However, it is unlikely that these camps were constructed in the first instance to supervise mining in the locality; they were clearly intended as Roman legionary camps during the campaigns in this area of Q. Servilius Caepio against Viriathus and Q. Caecilius Metellus Pius against Sertorius respectively. (70) The relatively thorough excavations at the military camp on the outskirts of modern Cáceres allow its chronological horizons to be established with some certainty : viz. that it went out of occupation c. 80 B.C. If it were intended as a fort to guard the mining operations, its period of occupation would have extended into the Principate. (71) Similarly Castra Caepiana on the Setúbal peninsula may have fulfilled some supervisory function for the gold extraction along the Tagus, but since there has been no excavation of this site to allow greater precision over its period of occupation, it is impossible to prove whether it should be connected with Roman exploitation of the Tagus.

A military presence can also be detected in the upper Tagus valley in the area around Idanha. An inscription records a _praefectus equitatae alae I singularium civium romanorum_. (72) It has plausibly been suggested that this _ala_ was resident in the urban centre of the Igaeditani to oversee the important gold mines of the area. (73)

A supervisory force was necessary in mining areas to provide protection for the minerals (both when they were waiting to be processed and when they were being transported in ingot form) and also to keep the large labour force (partly criminal) under control. The _Lex Metallis Dicta_ laid down the penalty for stealing ore when it was waiting to be processed (74) and Antoninus Pius also instituted regulations for the punishment of those convicted of stealing gold or silver from imperially owned mines. (75) Soldiers would have helped to enforce such regulations. Secondly, the large labour forces needed in larger mining communities, being of mixed slave, free and convict personnel, would have been a natural focus for unrest, which soldiers would have been on hand to control. (76)

Finally, soldiers also provided technical expertise and assistance not only in mining itself (for example, in the preparation or repair of shafts and galleries), but also in keeping open communications' links with the outside world, vital for the import of supplies and export of metals (by, for example, constructing and maintaining roads and bridges). (77)

ii. Mining Personnel :

Landowners, Lessees and Labourers

It now remains to consider the social status and financial position of the individuals involved in Roman mining in Lusitania. In my view, there were three general levels of involvement : first, those who owned strains of mineral on their own estates; secondly, those who leased the contracts to exploit the imperially-owned mines (conductores) and thirdly, the labourers who actually extracted the ore from the ground and later processed it.

Again the legal texts from Vipasca are instructive. A prospective lessee of a silver shaft had to bid at an auction for the right to work it, in return for which he was allowed to keep half of the ore extracted. The other half belonged to the fisc. But if he paid an additional 4,000 sesterces to the fisc, he was entitled to the other half of the produce as well. (78) For comparison, this same amount was bequeathed at Abellinum in Italy for the building of a portico; it was, therefore, a considerable sum. (79) We have no indication of how much the lessee would expect to pay at auction for the right to exploit minerals at Vipasca.

That there were difficulties involved for some in the raising of this capital is suggested by a number of points in the legal texts. First, twenty-five days were granted for the raising of the capital (80) and, secondly, the right to form associations (societates) was allowed. (81) Moreover, provisions were laid down to prevent significant interruptions in the exploitation of a source of metal. This may in part imply that periods of inactivity might be caused by difficulties in raising capital for the mining operations. (82) As an additional expense, lessees were responsible for the upkeep of their own shafts. (83)

All this would suggest that a lessee would have had to have possessed a substantial amount of capital to become involved in the extraction of minerals, although the right to form associations might have helped to encourage the participation of men of lesser means.

(84) It has been recently noted that shafts seem often to have been sunk in pairs, which would give some archaeological confirmation of co-operative ventures. (85)

Moreover, associations were common in mining operations elsewhere in the Iberian peninsula : hence the many lead ingots with stamps bearing the names of societates : for example, the societas Mont. Argent. Ilucro. (86) or the series of ingots discovered at Marseillan, Hérault attesting a Soc(ietas) M. (et) C. Pontilienorum M.f. (87) or the Soc. L. Gargili T.f. et M. Laetili M.l. (88)

Most of the shafts so far revealed at Aljustrel are of small dimensions (1m x 1m) and many superficial open-cast workings have also been discovered. (89) This all suggests a lot of small-scale operations. Although total production at Vipasca was clearly large (as the existence of special procurator to supervise operations suggests), it was broken up into many small units of production. It is thus misleading to think of the Vipasca mines (and other large Roman mining sites) in modernist terms as "large-scale" or "state-run". They did not consist of a single mining operation, controlled directly by the mining procurator; production was devolved into much smaller concerns. It has even been suggested recently that the mining procurator was not permanently resident at Vipasca, but possibly spent some time at the provincial capital, Augusta Emerita, in the office of the provincial procurator, to whom he was subordinate in the overall financial administration of province. (90)

The difficulty and cost of mining operations were circumscribing factors that allowed only people who had a solid basis of wealth to invest in mining. A recent account of Roman silver mining will serve to illustrate the ramifications in terms of the investment, equipment, fuel and organization of production needed for mining operations : (91)

> "(The production of just one tonne of silver) involved digging up roughly 100,000 tonnes of rock with iron and stone picks, hammers, chisels and gads or by fire setting; this rock had to be dragged to the surface in baskets by ropes and human muscle. In deep workings ..... this labour would involve the use of tonnes of illuminating oil. Once the ore was brought to the surface, it was broken up by hammers, picked

> over by hand, ground and washed. Ten thousand tonnes of trees provided enough charcoal (between 500 and 2,000 tonnes) to smelt the ore at the necessary temperature (1,000 degrees C.). The bye-product was 400 tonnes of lead and lots of slag which had to be carted away. Often the silver was still impure and had to be washed and resmelted to produce refined silver."

It is unlikely that fortunes could be made initially from mining alone : the scale of initial investment must have meant that a basic source of wealth was necessary for those who wanted to become involved in mining. An obvious source of this wealth would have been the land.

This has been convincingly demonstrated for the operation of the Attic silver mines at Laurion. Leases survive which show that the status of lessees varied, but that certain individuals of known wealth and landed property were strikingly active in both the mines and the processing plants, including a number of trierarchs, who were drawn from the richest class of Athenian citizens. (92)

Similarly there may have been a close connexion between landed wealth and the extraction of precious metals in Lusitania. This is possibly suggested by the striking coincidence between the location of mines and some rural villa sites, indicative of a relatively high level of wealth. The evidence is best summarized in tabular form (Table 4.3).

TABLE 4.3 ROMAN VILLAS LOCATED IN PROXIMITY TO MINES

VILLA (*)	MINE(S)	REF.+
A. CONVENTUS PACENSIS		
1. Monte do Farrobo, Aljustrel (PS 01)	Vipasca	P 33
2. Tapada da Fonte Soeiro, Alandroal (PS 23)	Bugalho	P 41
3. Nossa Senhora de Tourega (PS 26)	Serra dos Monges Nogueirinho	P 39 P 40
(& marble quarries at São Brissos : AEA, 1977-8, 179)		
4. Vila Viçosa (PS 29)	Vila Viçosa	P 42
(& marble quarries at Estremoz : AEA, 1977-8, 179)		
5. Montinho dos Laranjeiras, Alcoutim (PS 31)	Various mines in concelho of Alcoutim	P 21-29
6. Alamo, Alcoutim (PS 32)		
7. Amendoal, Amendoal (PS 33)	slag at Amendoal	P Slag 15
8. Milreu, Estói (PS 36)	slag at Milreu	P Slag 13

B. CONVENTUS SCALLABITANUS

9. Citania da Rosa, Ferro (PC 01)	Ferro, Covilhã	S 30
	(3 possible villa sites here)	*Estudos de Castelo Branco*, 1974, 67
10. Vale da Caradinha, Meimoa (PC 03)	Meimoa	S 26
11. Arneiro, Arnal (PC 06)	Arneiro	S 02
12. Pedrogão, Povoa de Cos (PS 08)	foundry near villa	S 01
13. Telhões, Pero Martins (*AP*, 1974-7, 211)	Telhões	S 33

C. CONVENTUS EMERITENSIS

14. Finca de Berry, Albuquerque (BA 01)	Tres Arroyos	E 06
15. Leoncillo, Villar del Rey (BA 32)	Leoncillo	E 07

NOTES : * All references to J.-G. Gorges, *Les villas hispano-romaines*, Paris, 1979, except where stated.

\+ All references to Catalogue of Mines (Appendix I), except where stated.

This correlation between mines and villa sites might be simply the result of chance, and reflect no actual connexion between landowning and mining in Roman Lusitania. But it remains a distinct possibility that an owner of land in proximity to mines would have been a likely investor in the exploitation of those mines. At Telhões, Pero Martins (No. 13) a possible Roman villa site has been discovered only five hundred metres from Roman tin mines.

Another problem that arises with this evidence is that of chronology, i.e. were the mines of a given area in exploitation at the same time as the villa was in occupation ? Comments have already been made about the difficulty of dating mines and the periods of Roman exploitation. (93) Chronological problems abound also in the study of Roman villas of Lusitania. Gorges has done a useful service not only in cataloguing all known villa sites, even if a supplement is now needed to his work, but also in outlining some of the problems over dating such sites. (94) The main problem remains that only in very rare cases has a rural villa site been completely excavated. Thus we are often left with merely the later phases of a villa. This factor has led to what may well be an erroneous picture of the rural economy of Lusitania emerging, viz. that the Romanized villa was a phenomenon of the later Empire in this province. The recent Franco-Portuguese excavations at São Cucufate, Vidigueira (PS 22) throws the entire gamut of problems into high relief. Previously it was believed that this site was established in the third century A.D.; but it can now be shown to have been occupied from the first century A.D. (95) It is therefore almost impossible to establish any chronological relationship between the villas and mines listed in Table 4.3. The only chronological correspondence that can be demonstrated is for No. 1 : an inscription from the villa at Monte do Farrobo has been dated to the first century A.D., when the mines of Vipasca were, as has been shown, already in operation. It is thus very possible that M. Valerius Rufus and his sons involved themselves as lessees in the mining that took place close to their rural estate.

More secure evidence is provided by those sites where metallurgical installations have been discovered as annexes to the residential centre of a rural estate : for example, at Arneiro, Arnal (No. 11 in the Table), Pedrogão, Povoa de Cos (No. 12) or Leoncillo, Villar del Rey (No. 15). Even here it does not necessarily have to follow that the landowner had been involved in the mining of the metal that he was smelting. Furthermore, are these traces of metallurgy (mostly the presence of slag) to be seen to represent the existence of installations for the smelting of raw metal ore ? They may merely reflect kilns for the conversion of already smelted metal, especially iron, into domestic or agricultural implements, and thus have no connexion with local mining. However, this is the most pessimistic interpretation of the evidence such as it is ; and it does not exclude the hypothesis that given the proximity of mineral resources, and furthermore given the often generally poor conditions for agriculture in the vicinity of mines, the involvement of local landowners in some aspect of mining, ore-treatment or metallurgy is highly probable.

Two mechanisms for this connexion between landed wealth and mining are possible. Either the villa owner may have become wealthy as a result of his mining activity and then used some of his wealth to invest in land; or he may have derived his wealth first from the land, which he subsequently invested in mining operations. On either interpretation the involvement of some of the Lusitanian landowning elite in mining is hereby illustrated.

It was not only the extraction of minerals that required substantial investment, but also the smelting and processing of the extracted ore. The possible existence of smelting works on landed estates in Lusitania has already been considered. In the Athenian mining operations at Laurion, landowners similarly invested in industrial plants on their estates. (96)

The main outlay for smelting was the fuel needed to heat the foundry to the necessarily high temperatures. Charcoal constituted the main source for fuel in the Roman period and the main kind of wood available for charcoal in the south-west of the Iberian peninsula was the evergreen oak (Quercus ilex), although other varieties could also have been used, for example the cork oak (Quercus suber) and the scarlet oak (Quercus coccifera). (97) To provide some indication of the sort of orders of magnitude we are dealing with, some speculative calculations for annual charcoal requirements have been made for three large mines in the south-west of the Iberian peninsula : Vipasca and Serra da Caveira in Lusitania and Rio Tinto in neighbouring Baetica (Table 4.4).

TABLE 4.4

HYPOTHETICAL CHARCOAL REQUIREMENTS AT (A) VIPASCA (B) SERRA DA CAVEIRA (C) RIO TINTO

I. Some figures

i. 1 tonne of charcoal produced by 5 tonnes of timber

ii. 1 tonne of timber acquired from c. 3.33 oak trees (*Quercus ilex*)

iii. Smelting A (Cupreous pyrites):

1 tonne of slag produced by 1.833 tonnes of charcoal

iv. Smelting B (Jarositic earth):

1 tonne of slag produced by 1 tonne of charcoal

II. Calculation

i. $\dfrac{\text{slag}}{\text{period of Roman mining}}$ = slag per annum

ii. Method A : slag per annum x 1.833 = charcoal per annum

Method B : slag per annum x 1 = charcoal per annum

iii. charcoal per annum x 5 = tonnes of timber per annum

iv. timber per annum x 3.33 = rough number of trees (or tree equivalent) per annum

(A) VIPASCA

i. $\frac{3,000,000}{300}$ = 10,000 t slag per annum

ii. Method A : 10,000 x 1.833 = 18,330 t charcoal per annum

Method B : 10,000 x 1 = 10,000 t charcoal per annum

iii. A : 18,330 x 5 = 91,650 t timber per annum

B : 10,000 x 5 = 50,000 t timber per annum

iv. A : 91,650 x 3.33 = 305,194.5 trees per annum

B : 50,000 x 3.33 = 166,500 trees per annum

(B) SERRA DA CAVEIRA

i. $\frac{300,000}{300}$ = 1,000 t slag per annum

ii. A : 1,000 x 1.833 = 1,833 t charcoal per annum

B : 1,000 x 1 = 1,000 t charcoal per annum

iii. A : 1,833 x 5 = 9,165 t timber per annum

B : 1,000 x 5 = 5,000 t timber per annum

iv. A : 9,165 x 3.33 = 30,519 trees per annum

B : 5,000 x 3.33 = 16,650 trees per annum

(c) RIO TINTO

i.		$\frac{15,300,000}{400}$		= 38,250 t	slag per annum
ii.	A :	38,250 x 1.833	=	70,112 t	charcoal per annum
	B :	38,250 x 1	=	38,250 t	charcoal per annum
iii.	A :	70,112 x 5	=	350,561 t	timber per annum
	B :	38,250 x 5	=	191,250 t	timber per annum
iv.	A :	350,561 x 3.33	=	1,167,369	trees per annum
	B :	191,250 x 3.33	=	636,863	trees per annum

Source for calculation : L.U.Salkield, 'Ancient slags of the south-west of the Iberian peninsula' in La Mineria Hispana e Iberoamericana, León, 1970, I, 85-98.

These figures, approximate though they are, emphasise that an enormous quantity of fuel was needed for processing metals. Moreover, these figures are conservative estimates, minimal figures, since the main periods of operation at each of the mines considered were probably shorter than the figure used for the purpose of the calculation. Similarly there will have been some periods when activity was much greater than the average, thus increasing still further the need for fuel. Again, some of the ore extracted may well have been processed away from the mining zone itself. Thus the slag surviving from a mining site need only represent a proportion of the total ore extracted in the Roman period. Further timber would also have been needed for props, shafts, galleries and water-lifting wheels. A local landowner would have been better placed than others in that he could have produced at least some of the required timber on his own estate. (98)

The exhaustion of timber supplies has often been used as a partial explanation for the reduction in the scale of operations, or even for the termination of activity, at various mining sites. (99) Deforestation is, therefore, to be expected in the hinterland of mines. So the area around Sounion in Attica was a major source of timber for the Laurion silver mines, but quickly became denuded of trees. (100) Similarly in seventeenth century England the enormous consumption of timber for fuelling smelting furnaces led to critical shortages of supply in some areas. (101)

There were other ingredients required for smelting which would have had to have been imported into mining areas : for example, salt, which was used as a separating agent in the smelting of silver. (102) Again local landowners would have been the most likely to have had the necessary capital to acquire these materials from the traders who brought them to the mining areas. Furthermore, openings would thus have been created within the local economy for traders to supply such materials, another economic effect of the expansion of mining under the Romans.

From all this it is clear that substantial investment was necessary for mining and for smelting and processing. The Vipasca mines had their own foundries, which were leased to contractors (conductores) under leases separate from those that regulated mining. (103) A similar distinction existed at Laurion between the leases for the mines themselves and those for the processing plants (104) Another provision of the Lex Metalli Vipascensis implies that the expense and problems involved in smelting gave each foundry an especial value, if not exclusivity. For it was laid down that anyone who brought ore which had been extracted elsewhere to be smelted at Vipasca had to pay a special fee of one denarius per hundred pounds to the fisc. (105) This suggests that there were some small operators active in other (? non-imperial) mines in the vicinity, who had insufficient funds to provide for their own foundries. The richer landowners did have the necessary capital to run their own foundries on their own estates, as occurred, for example, at the villa at Pedrogão, Povoa de Cos (PC 08). (106)

Another technological problem that would have required substantial effort and investment to overcome was the drainage of the mining shafts. Flooding and water in the shafts was a constant problem. Although some drainage was effected by hand, various water-lifting devices have been discovered in Roman mines, all of which must have involved a substantial investment to set up : for example, the Archimedean screw found at Sotiel Coronada and Santa Barbara (Posadas) or the series of water-lifting wheels found at Rio Tinto and São Domingos. (107)

To date, no drainage machinery as such has been found at Aljustrel, but buckets of esparto grass have, as well as fittings from bronze vessels, and some bronze vessels themselves. These were used for the transfer of raw ore from the mine shaft to the surface, but may also have been used to clear any water that gathered in the shaft. (108) The workings at Vipasca are not generally deep, and so the need for machinery was not as pressing as in others mines in the peninsula. There is reference to drainage tunnels (cuniculi) in both copper and silver workings at Vipasca. (109) Since Vipasca consisted of many small workings, most of the mining was concerned with ores that lay near the surface, or even open-cast mining was practised. Especially in silver mines the best ore is just that near the surface, and furthermore it requires less sophisticated smelting procedures than lesser-quality deeper ore. Thus, as soon as the surface ores were exhausted at Vipasca, lessees probably moved on to open up new workings, rather than dig deeper in the same shaft, which would have involved a lesser return for a greater effort expended. In other words in silver mining the law of diminishing returns applies. (110)

Inscriptions from the Vipasca mines themselves and the surrounding territory suggest themselves as possible sources of information on the social makeup of a mining community or local landowners who might have become involved in the mining operations. The evidence is set out in Table 4.5.

TABLE 4.5

INDIVIDUALS ATTESTED EPIGRAPHICALLY FOR THE VIPASCA MINES AND ENVIRONS

A. VIPASCA

1.	T. Iunius	NRHDFE 1908, 306-310 IRCP 121
2.	P. Aelius Modestus	NRHDFE 1908, 311-312 AP 1967, 131; IRCP 134
3.	Modestianus (son of 2)	ibid.
4.	Laberia M.f. Coimia	AE 1978, 381 = IRCP 131
5.	L. ...cisicus Maxumus	AE 1978, 382 = IRCP 133
6.	M. Postumius L.f.	AE 1978, 384 = IRCP 136
7.	C. Anuius (or Anvius or Annius)	AE 1978, 378 = IRCP 123
8.	[Bo]uta Iuliae f.	AE 1978, 380 = IRCP 125
9.	Cn. Naeidius Rufus	AE 1978, 383 = IRCP 135
10.	L. Virius Calatius	unpublished; referred to by C. Domergue, op.cit. (n. 24), 34
11.	Palentia	Conimbriga, 1966, 7-104 : Grave 206
12.	Tancus	ibid. : Grave 213

13. Quintilia (or Quintilla)	unpubl.; referred to by C. Domergue, op.cit. (n. 24), 34

B. TERRITORY

i. Monte do Farrobo, Rio de Moinhos

14. M. Valerius Rufus	*Conimbriga*, 1959, 109-14; *IRCP* 137
15. M. Valerius Marcellus (son of 14)	ibid.
16. M. Valerius Marcellus (son of 15)	ibid.

ii. Panoias, Ourique

17. C. Iulius Bouti f. Letondo	*CIL* II 5790; *IRCP* 130

Since only the relatively wealthy could afford to erect inscriptions, those thus attested may well have had the necessary capital to invest in mining leases at Vipasca. Only one of those attested gives any indication of his status : T. Iunius (no.1), the sole individual whose name is preserved of a group of *coloni* who set up a statue to a procurator of the Vipasca mines. It is a pity that the names of his colleagues have become illegible on the inscription. Who were these *coloni* ? It is possible that they were lessees of contracts to exploit mining shafts. But the usual term for these was *conductores*, as the Vipasca legal texts show. *Coloni* are referred to in the *Lex Metallis Dicta*, but this is a document of general application, and not just for the Vipasca mines where it was found. I would argue that these *coloni* were workers in the mines who were either working for the lessee (*conductor*) or who had sub-let part of a contract from a *conductor*. This process of sub-letting is well known in other legal contexts. (111)

P. Aelius Modestus (no. 2) was certainly involved at Vipasca, since he is attested burying his son in the cemetery of Valdoca, but it is impossible to be certain of his exact role. Most of the individuals attested are freeborn, with a number of indigenous names present : e.g. L. Virius Calatius (no. 10) or C. Iulius Bouti f. Letondo (no. 17), the latter still maintaining the Celtic style of filiation, even though he has taken a Romanized *praenomen* and *nomen*. The status of Palentia (no. 11) and Tancus (no. 12) is uncertain, although their names suggest that there were of indigenous stock. Their names appear not on inscriptions, but as graffiti scratched on coarseware pots from the Valdoca cemetery. The elegance of the inscription recording the M. Valerii (nos. 14, 15, 16) and its discovery near a rural villa site suggest that they were a family of some substance, who would have possessed the necessary wealth to invest in mining at Vipasca. (112) The same would apply of the other landowners in the hinterland of Pax Iulia, from where a relatively high number of villa sites have to date been discovered. (113) But in sum these inscriptions do not get us much closer to the people who were involved as lessees at Vipasca, since (with the exception of no. 1) we can never be sure that the individuals attested were involved with mining.

As regards the question of the labour force in the mines of Lusitania, some remarks of Finley will serve as a starting-point : (114)

> "Mining has always been an exceptional occupation, reserved .... for the depressed sectors of the population, slaves where they were available, free men whose freedom is fragile and easily encroached upon where no slavery exists. Throughout antiquity free miners were a negligible element."

In imperially owned gold and silver mines prisoners-of-war or convicts provided an obvious source of labour. (115) It has been suggested that there was some forcible transfer of Lusitanians to provide labour in the mines of the *conventus Pacensis*. (116) The mining settlement at Vipasca has recently been excavated and has revealed a series of houses of low-quality fabrication. The artefacts from there are few in number and suggestive of a low level of wealth. Similarly, the grave-goods from the cemetery of Valdoca would suggest that the cemetery was in general used for individuals of a fairly low financial standing. (117)

The legal texts from Vipasca are more promising. From them it is clear that conductores used servile labour to work down the mining shafts. For a number of offences have differentiated penalties prescribed for slaves and freeborn. (118) One of the penalties concerns prospecting too close to a drainage gallery, which would imply that slaves were involved in the actual extraction of metals. (119) It is also clear that slaves (and possibly freedmen) were used in smelting and processing. (120)

However, the use of slave labour did not exclude other forms of labour. For hired free workers (mercenarii) were also employed at Vipasca. (121) The use of mercenarii can be paralleled from the Dacian gold mines, from where a contract between a free conductor and a free (if illiterate) labourer. (122) Just as free, hired labour is now seen to have been a integral part of Roman agriculture, so its role in mining should not be ignored. (123)

iii. Conclusion

The impact of the Roman conquest on the mining and metal industries of Britain has recently been investigated. There was clearly an expansion under the Roman Empire, not as a result of the introduction of new techniques, but rather in the form of an increase in the number of productive units. The same is true of the mines of the Iberian peninsula and has been clearly demonstrated for Vipasca and Rio Tinto. The increase in production, it has been noted, was stimulated in three ways : by the requirements of the Roman army, by the increase in urbanization (metal nails, hinges and other building materials were a feature of Roman, as opposed to native British, building techniques) and by the adoption of a monetized economy on a far wider scale than before. (124)

Such factors were equally important in the expansion of the mining industry in Lusitania. The expansion was underpinned by fundamental changes in the social and economic organization of the province that came with integration within the Roman Empire and with pax romana. Furthermore the construction and maintenance of the road network and the regularized use of river and sea routes allowed the metals to be transported with greater facility from the mines to production centres to be converted into arms and armour, agricultural and mining tools, building materials or, perhaps most important of all, into coinage.

The very fact that mining in the area declined rapidly after the departure of the Romans in the wake of the Suevic and then Visigothic occupation of the Iberian peninsula is eloquent testimony for the importance of the Roman occupation as a crucial stimulus to the development of mining in the area. This is not to say that there had not been mining before the arrival of the Romans, or that there was not any after their departure. But the fact that Lusitania was now part of a wider political unit, the Roman Empire, had an important effect not only on the local society, but also on the local economy.

NOTES : CHAPTER FOUR

1. For commentaries on these texts see A. d'Ors, Epigrafia juridica de España romana, Madrid, 1953, 80-133; D. Flach, 'Die Berkwerksordnungen von Vipasca', Chiron, 9, 1979, 399-448; C. Domergue, La mine antique d'Aljustrel (Portugal) et les tables de bronze de Vipasca, Paris, 1983, all with extensive bibliographies of earlier work.

2. O. Davies, Roman Mines in Europe, Oxford, 1935, 13; A. d'Ors, 'La condición juridica del suelo en las provincias de Hispania' in I diritti locali nelle province romane con particolare riguardo alle condizioni giuridiche del suolo (= Accad. Naz. dei Lincei, 194), Rome, 1974, 253-68, esp. 267; for this in Danubian provinces see S. Dusanic, 'Roman mining in the Danubian provinces', ANRW, II.6, Berlin/New York, 1977, 53-94, esp. 89-90.

3. For vicus : VIP A, 5.i, line 37; for territory : "intra fines metalli Vipascensis" (VIP A, 1.i, line 2; 2.i, line 10), "inve territoris eius (sc. metalli Vipascensis)" (VIP A, 5.i, lines 37-38).

4. Official notices displayed in forum : VIP B, 6.ii, line 15; for an example of a statue see references in Chapter Three, note 67.

5. So, for example, the tesserae marked "S.C." from the mines of Castulo in Baetica : M.P. Garcia y Bellido, Las monedas de Castulo en escritura indigena : historia numismatica de una ciudad minera, Madrid, 1982; or the examples from the mine of Plasenzuela (E 10) now in the Museo Arqueologico Nacional in Madrid; or those preserved in the Archivo Municipal in Seville, bearing a representation of a miner and the legend CELTE, plausibly connected with the mines at Celti (Penaflor) in the Guadalquivir valley : A. Blanco Freijeiro & J.M. Luzon, 'Mineros antiguos españoles', AEA, 39, 1966, 73-88, esp. 87 & figs. 6 & 7 (p. 80). For the concept in general see O. Davies, op.cit. (n. 2), 13.

6. Cornish tin mines : kind information of International Tin Research Institute, Greenford (Middx.); Yorkshire collieries : J. Benson, British Coalminers in the Nineteenth Century : a social history, Dublin, 1980, 88-93.

7. Auctioneering : VIP A, 2, lines 10-18; baths : VIP A, 3, lines 19-31.

8. H. Cockle, 'Pottery manufacture in Roman Egypt : a new papyrus', JRS, 71, 1981, 87-97.

9. ibid., 87, text lines 13-16.

10. VIP A, 3.ix, lines 28-29; 3.xii, line 30.

11. Similarly in the Derbyshire lead mines in the seventeenth century the right to gather wood for fuel was granted only to certain individuals : see R.J. Hopper, 'The Laurion mines : a reconsideration', ABSA, 63, 1968, 293-326, esp. 305.

12. See R. Meiggs, Trees and Timber in the Ancient Mediterranean World, Oxford, 1982, 329 citing Hyginus 114.3 (ed. Lachmann) for public building and Frontinus 55.4 (ed. Lachmann) for baths.

13. CIL II 3361 = ILS 5688; for the figure of 37 hectares see R.P. Duncan-Jones, 'The procurator as civic benefactor', JRS, 64, 1974, 79-85, esp. 85, note 53.

14. VIP A, 8, line 55. Exemption from tax (instituted by Vespasian) : AE, 1936, 128 = FIRA I 73.

15. J. & A. de Alarcão, 'O espolio da necropole luso-romana de Valdoca (Aljustrel)', Conimbriga, 5, 1966, 7-104.

16. C.C. Patterson, 'Silver stocks and losses in ancient and medieval times', EHR, ii, 25, 1972, 205-235.

17. 3.2.10 (148)

18. J.S. Richardson, 'The Spanish mines and the development of provincial taxation in the second century B.C.', JRS, 66, 1976, 139-52, esp. 146.

19. The details are conveniently set out by Richardson, art.cit. (n. 18), 152.

20. For the economic needs created by the mining at Potosi see J. de Mesa & T. Gisbert, 'Oruro : origem de una villa minera' in La Mineria Hispana e Iberoamericana, León, 1970, I, 559-590, esp. 580-90. For Laurion in Attica see Xen. Por. 4.49.

21. Strabo 3.2.8 (146); Plin. NH 33.21.67. In general see T.A. Rickard, 'The mining of the Romans in Spain', JRS, 18, 1928, 129-43, esp. 139.

22. For the problems of supplying Rio Tinto see S.G. Checkland, The Mines of Tharsis : Roman, French and British Enterprise in Spain, London, 1967, 50.

23. So in sixteenth and seventeenth century Mexico the newly worked mines attracted immigrants who provided supplies and services for the mining communities : F. Chevalier, Land and Society in Colonial Mexico (tr. A. Eustis), Berkeley, 1963, esp. 38-41. See also P.J. Bakewell, Silver Mining and Society in Colonial Mexico : Zacatecas 1546-1700 (= Cambridge Latin American Studies, 15), Cambridge, 1971.

24. Dig. 48.19.8.4; see further F. Millar, 'Condemnation to hard labour in the Roman Empire, from the Julio-Claudians to Constantine', PBSR, 52, 1984, 124-47, esp. 137-43.

25. S. Dusanic, art.cit. (n. 2), 93.

26. Transportation : 3.2.3 (142); metals : 3.1.6 (139)

27. M. Maia, 'Fortalezas romanas do sul de Portugal', *Zephyrus*, 28-29, 1978, 279-85.

28. See Chapter Three, p. 47 & note 56.

29. *AEA*, 32, 1959, 167-68.

30. *CIL* II 967

31. J.M. Blazquez, 'Fuentes literarias griegas y romanas referentes a las explotaciones mineras de la Hispania romana' in *La Mineria Hispana e Iberoamericana*, León, 1970, I, 117-50, esp. 140.

32. *CIL* II 971. On mining around Serpa : J. de Alarcão, *Portugal romano*, (3rd ed.), Lisbon, 1983, 131, 134-5 : especially the major copper/silver mines at São Domingos.

33. *CIL* II 964. On location of Turobriga : L. Garcia Iglesias, 'La Beturia : un problema geografico de la Hispania antigua', *AEA*, 44, 1971, 86-108, esp. 98-99.

34. J.M. Luzon & D. Ruiz Mata, 'El poblado minero de Rio Tinto', *Habis*, 1, 1970, 125-38, esp. 137-8.

35. J.M. Luzon, 'Lucernas romanas de Riotinto', *AEA*, 40, 1967, 138-150, esp. 141.

36. C. Belchoir, 'Duas notas sobre lucernas da necropole romana de Valdoca (Aljustrel)', *Conimbriga*, 9, 1970, 75-79. There is a problem over the dates of these various lamps. The Aljustrel examples can all be firmly dated to the first century A.D. by the artefacts found with them in graves. Luzon, art.cit. (n. 35), dated the Rio Tinto lamps to the second century on the basis of coins, which would mean that we are dealing with two different periods. However, it is not certain that these coins were found in the same archaeological context as the lamps, which makes them useless as dating criteria (see Belchoir, art.cit., p. 77). Even if we are not dealing with

the same potter, we are probably dealing with the same pottery firm.

37. On branch workshops in the lamp industry see W.V. Harris, 'Roman terracotta lamps : the organization of an industry', JRS, 70, 1980, 126-145.

38. For Paternus see O. da Veiga Ferreira, 'Ara votiva de Lousa', Rev.Guim., 62, 1952, 192-195. For all other references see Table 4.1.

39. D.F. de Almeida, Egitania : Arqueologia e Historia, Lisbon, 1956. On immigration in the province in general see J.C. Edmondson, Economy and Society of Roman Lusitania under the Principate (unpublished Ph.D. thesis), Cambridge, 1984, Chapter III.

40. Kind information of Professor Alarcão, on the basis of his forthcoming gazeteer of Roman archaeological sites in Portugal.

41. CIL II 5132

42. A theatre at Madauros cost 375,000 sesterces, a temple of Magna Mater at Lepcis Magna 200,000 sesterces : R. P. Duncan-Jones, The Economy of the Roman Empire : quantitative studies, Cambridge, 1974, 90-91.

43. On emigration from Clunia see G. Fabre, 'Le tissu urbain dans le nord-ouest de la péninsule ibérique', Latomus, 29, 1970, 314-39, esp. 318-23, 332-6.

44. IRCP 339; J.C. Edmondson, 'Mithras at Pax Iulia : a re-examination', Conimbriga, 23, 1984, 69-86.

45. IRCP 122; R.M. Varela Gomes, 'Uma estela epigrafada da Herdade dos Franciscos (Garvão)', Conimbriga, 23, 1984, 43-54.

46. VIP B, 1, line 1

47. See Chapter Three, note 67.

48. AE 1933, 273; C. Habicht, Die Inschriften des Asklepeions (= Altertumer den Pergamon, VIII.3), Berlin, 1969, 88-92, no. 44.

49. CIL II 2598

50. CIL II 1179 = ILS 1591

51. ILS 1592

52. ILS 276

53. For Saturninus : P. le Roux, 'Procurator affranchi in Hispania : Saturninus et l'activité minière', MDAI(M), 26, 1985, 218-33; Silvanus : A. Tranoy, La Galice romaine : recherches sur le nord-ouest de la péninsule ibérique dans l'antiquité (Publ. du Centre Pierre Paris, 7), Paris, 1981, 183-4.

54. S. Dusanic, art.cit. (n. 2), 87.

55. C. Domergue, 'Introduction à l'étude des mines d'or dans le nord-ouest de la péninsule ibérique' in Legio VII Gemina, Leon, 1970, 255-86, esp. 279.

56. CT 1.32.5 : procuratores metallorum ...... soliti ex curialibus ordinari

57. O. Davies, op.cit. (n. 2), 9; J.F. Healy, Mining and Metallurgy in the Greek and Roman World, London, 1978, 131.

58. See Chapter Three, notes 43 & 50.

59. Diod. Sic. 3.13.1

60. For photo see A. Blanco Freijeiro & J.M. Luzon, art.cit. (n. 5), 88 & fig. 10. Another relief depicting a child from the mines of Castulo is preserved in the Museo Arqueologico Nacional in Madrid.

61. Tac. Ann. 11.20

62. R.F.J. Jones, 'The Roman military occupation of north-west Spain', JRS, 66, 1976, 44-66.

63. Pumpsaint : P.R. Lewis & G.D.B. Jones, 'The Dolaucothi gold mines, I : the surface evidence', Ant. J., 49, 1969, 244-72. Charterhouse : see plan at Britannia, 2, 1971, 277; Mendip lead pigs : CIL VII 1202; CIL XIII 3491 = ILS 8709.

64. CIL III 25

65. S. Dusanic, art.cit. (n. 2), 73.

66. J. Maluquer de Motes, 'Excavaciones arqueologicas en el castro de "Las Merchanas" (Lumbrales, Salamanca)', Pyrenae, 4, 1968, 101-28, esp. 119.

67. M. Maia, art. cit. (n. 27); for an update on more recent excavations see M. & M. Maia, Informação arq., 4, 1981, 45-6.

68. M. Maia, art. cit. (n. 27), 281.

69. Plin. NH 4.22.117

70. A. Tovar, Iberische Landeskunde II.2. Lusitanien, Baden-Baden, 1976, 237-8 (Castra Caecilia), 238 (Castra Servilia).

71. G. Ulbert, Cáceres el Viejo. Ein spätrepublikanisches Legionslager in Spanisch Extremadura (Madrider Beitrage, 11), Mainz am Rhein, 1984, 192-4, 202-4 for dating : camp goes out of occupation c. 80 B.C.

72. AE 1967, 145 = HAEp 1077; P. le Roux, L'armée romaine et l'organisation des provinces ibériques d' Auguste à l'invasion de 409, (Publ. du Centre Pierre Paris, 8) Paris, 1982, 270-74.

73. J. de Alarcão, op. cit. (n. 32), 108.

74. VIP B, 9. ii - 10, lines 25-29

75. Dig. 48.13.8.1

76. G.D.B. Jones, 'The Roman mines at Rio Tinto', JRS, 70, 1980, 146-65, esp. 162.

77. P. le Roux, 'L'armée de la péninsule ibérique et la vie économique sous le Haut-Empire romain' in Armées et fiscalité dans le monde antique (Colloque du CNRS, Paris, 1976), Paris, 1977, 341-372, esp. 363-364.

78. VIP B, 2, lines 6-7; cfr. comments of C. Domergue, op.cit. (n. 1), 123-37.

79. CIL X 1136 and R.P. Duncan-Jones, op.cit. (n. 42), 161. Compare 3,600 sesterces for a portico at Mediolanum : Duncan-Jones, loc. cit.

80. VIP B, 4, lines 10-11

81. VIP B, 6 - 8, lines 14-24

82. VIP B, 4 - 5, lines 10-12

83. VIP B, 11, lines 30-31

84. Dig. 3.4.1 for the right to form societates in gold and silver mines.

85. C. Domergue, op.cit. (n. 1), 15-16.

86. *CIL* XV 7916

87. F. Laubenheimer & H. Gallet de Santerre, *Recherches sur les lingots de cuivre et de plomb d'époque romaine dans les regions de Languedoc - Roussillon et de Provence - Corse*, Paris, 1973, 140-145.

88. ibid. 136-138 : an interesting example of a business enterprise conducted by a freeborn individual in association with a freedman, on which see now P. Garnsey, 'Independent freedmen and the economy of Roman Italy under the Principate', *Klio*, 63, 1981, 359-371, esp. 365.

89. R.F. d'Andrade, 'A lavra romana das minas de Algares e na herdade do Montinho' *Actas do I Congresso Nacional de Arqueologia, Lisboa, 1958* Lisbon, II, 1970, 273-283, esp. 275; A. Viana, O. da Veiga Ferreira & R.F. d'Andrade, 'Exploração das minas de Aljustrel pelos Romanos', *Arquivo de Beja*, 13, 1956, 3-20, esp. 8.

90. P. le Roux, art. cit. (n. 53), 230-1.

91. K. Hopkins, 'Economic growth and towns in classical antiquity' in P. Abrams & E.A. Wrigley (ed.), *Towns in Societies*, Cambridge, 1978, 35-77, esp. 56.

92. R.G. Osborne, *Demos : the Discovery of Classical Attika*, Cambridge, 1985, 111-26; R.J. Hopper, 'The Attic silver mines in the fourth century B.C.', *ABSA*, 48, 1953, 200-254, esp. 246.

93. v. supra, p. 27-8.

94. J.-G. Gorges, *Les villas hispano-romaines : inventaire et problématique archéologiques* (Publ. du Centre Pierre Paris, 4), Paris, 1979.

95. J. de Alarcão, 'A vila romana de S. Cucufate', Arqueologia (Porto), 3, 1981, 117-21.

96. See R.G. Osborne, op. cit. (n. 92), 112-5.

97. Quercus ilex : L.U. Salkield, 'Ancient slags of the south-west of the Iberian penisula' in La Mineria Hispana e Iberoamericana, León, 1970, I, 85-98, esp. 94. Quercus suber and coccifera are both known from the area in the Roman period : see A. Schulten, Iberische Landeskunde : Geographie des antiken Spaniens, Strasbourg, 1955, 526-527 with references.

98. R. Meiggs, op.cit. (n. 12), 380.

99. For example at Rio Tinto : G.D.B. Jones, art.cit. (n. 76), 161.

100. R.J. Hopper, art. cit. (n. 11), esp. 310.

101. R. Meiggs, op.cit. (n. 12), 185, 379.

102. R.J. Forbes, Studies in Ancient Technology, III, Leiden, 1955, 174.

103. VIP A, 7, lines 46-56

104. See R.J. Hopper, art.cit. (n. 92).

105. VIP A, 7.ii, lines 50-51.

106. See J.-G. Gorges, op. cit. (n. 94), 461.

107. In general see O. Davies, 'Ancient Roman and medieval mining techniques', Trans.Inst.Min.Met., 43, 1934, 3-54, esp. 21-22; J. G. Landels, Engineering in the Ancient World, London, 1978, 58-83. For the examples from the Iberian peninsula see J. M. Luzon Nogue, 'Los sistemas de desagüe en minas romanas del

suroeste peninsular', AEA, 41, 1968, 101-120; C. Domergue, 'L'eau dans les mines d'or romaines du nord-ouest de l'Espagne' in P. Louis (ed.), L'homme et l'eau en Mediterranée et au Proche Orient. III : L'eau dans les techniques (= Travaux de la Maison de l'Orient, 11), Lyon, 1986, 109-119. Diod.Sic. 5.37.3 refers to the Archimidean screw.

108. C. Domergue, op. cit. (n. 1), 38.

109. VIP B, 14 - 15, lines 36-39.

110. This equally applied in later Spanish American silver mines : D. Brading & H. Cross, 'Colonial silver mining : Mexico and Peru', Hispanic American Historical Review, 52, 1972, 545-79, esp. 554-5.

111. J.A. Crook, Law and Life at Rome, London, 1967, 154.

112. R.F. d'Andrade, 'Uma lapide romana do Monte do Farrobo - Rio de Moinhos', Conimbriga, 1, 1959, 109-114.

113. For villas see J.-G. Gorges, op. cit. (n. 94), 471-7; for landowners of the area interested in mining see J. de Alarcão, 'Sobre a economia rural do Alentejo na epoca romana', Conimbriga, 15, 1976, 5-44, esp. 19-20.

114. M.I. Finley, The Ancient Economy, London, 1973, 72-3.

115. For this in north-west Spain see C. Domergue & G. Herail, Mines d'or romaines d'Espagne : le district de la Valduerna (León) : étude géomorphologique et archéologique, Toulouse, 1978, 282.

116. v. supra (p. 61 & note 26).

117. R. Parreira, arts. cit. (Chapter III, n. 66), J. & A. Alarcão, art.cit. (n. 15).

118. For example, for stealing ore : VIP B, 10, lines 27-29.

119. VIP B, 17, lines 41-3.

120. Slaves : VIP A, 7.i, line 49; freedmen : if the restoration of VIP A, 7.v, line 55 is correct as follows : "ex[cipiuntur servi et liberti].

121. VIP A, 7.i, line 49.

122. A. Berger, 'A labor contract of A.D. 164', C.Ph., 43, 1948, 231-242.

123. P. Garnsey, 'Non-slave labour in the Roman world' in P. Garnsey (ed.), Non-Slave Labour in the Greco-Roman World (= Cambridge Philological Society, Suppl. Vol. 6), Cambridge, 1980, 34-47.

124. W.H. Manning, 'The native and Roman contribution to the development of metal industries in Britain' in B.C. Burnham & H.B. Johnson (ed.), Invasion and Response : the case of Roman Britain (= BAR Brit.Ser. 73), Oxford, 1979, 111-121. See also M.I. Finley, 'Technological innovation and economic progress in the ancient world', EHR, ii, 18, 1965, 29-45, esp. 30 for the lack of Roman technological innovation in mining.

CHAPTER FIVE

GARUM PRODUCTION

i. Introduction

It is possible to draw together a body of literary, archaeological and numismatic evidence to attest the presence of fishing off the Lusitanian coast in the Roman period. The area was noted not only for its tunny and other pelagic species, but also for its oysters and shellfish. The Atlantic was reputed to be richer in fish than the Mediterranean. (1) Fish appear on the local coinages of seaboard towns (Baesuris, Ossonoba) and river ports (Salacia and Myrtilis). (2) Roman fishing equipment (fish-hooks, needles for net-making and ceramic and lead net weights) has been discovered in abundance along the sea-coast. (3) The essentially qualitative nature of this evidence, however, does not allow an assessment to be made of the importance of fishing to the Lusitanian economy. It also reveals little of the historical development of the activity.

Fishing could only have become economically significant, as in more recent times, if the catch could have been successfully processed and thus preserved. Processing can take on several forms : pickling in brine, curing, smoking, drying or conversion into a fish paste or sauce. The first three methods require the whole flesh of the fish, the last need only involve the scraps and the intestines of the fish; fish pastes/sauces are economically attractive because they are produced from parts of the fish that would otherwise go to waste. The economic importance of processing fish in modern times has been summed up as follows : (4)

> "In the first place, it [processing] serves to preserve a very perishable product; this

allows it to be used for a longer time after catching and to be transported over greater distances and so widens the market. Secondly, processing reduces the weight of the catch and reduces effective transport costs; and in the third place, there is direct economic importance in processing in the value that is added to the product."

In less developed societies processing was just as prevalent. Salted, dried and smoked fish all played a greater role in the medieval Portuguese economy than fresh fish, as they still do in the fishing communities of the Far East. (5) Firth isolated the advantages to be gained for the Malay fishermen in processing their catch : their products could be transported greater distances to markets, they helped deal with temporary surpluses of fish, they could be held in reserve against bad weather or seasons when fish were scarce and their preparation employed large numbers of people (notably the dependents of the fishermen themselves). (6)

The economic advantages to be gained from processing fish are quite clear from these modern accounts. But for those who controlled production the social benefits were just as important. Because of the diversity of activities involved, the production processes were labour-intensive; those who controlled production had the power to offer employment to a relatively large proportion of the local population at the peak fishing season, giving the producers an important patronage role. Furthermore, the fact that the product could easily be stored/hoarded without losing its nutrient value meant that the producers could regulate access to a foodstuff, the demand for which would have been drastically enhanced in times of general food shortage. By handing it out to the local populace at just such a moment of shortage the producers would have enhanced their social, and political prestige; or by releasing it onto the market at an inflated price, he could have reaped large financial rewards.

It is, therefore, the aim of this section to investigate the role of fish processing within the Roman economy of Lusitania. The discussion will consist of two complementary parts : first, an investigation of the organization of production and, secondly, of the diffusion and exchange of these products throughout the Mediterranean. However, it is first necessary to make some evaluation of the demand for processed fish and fish sauces in the Roman world.

ii. The Demand for Processed Fish

In the Roman world fish could be processed in two basic ways : the flesh could be cut up and salted (= salsamenta) or the leftovers (and/or small fry) could be macerated with salt to form various liquid fish-sauces. (7) The importance and ubiquity of Roman fish sauces is immediately suggested by the range of terms used to describe them : garum, liquamen, muria, hallec (sometimes referred to as allec, hallex or allex) and even possibly the more obscure lymphatum. (8) Garum also was mixed with other liquids to form still more condiments : oleogarum (garum mixed with oil), oenogarum (with wine), oxigarum (with vinegar) and, possibly, laccatum (with a herbal liquid). (9) There were different grades of the product, ranging from the expensive garum sociorum from New Carthage, a luxury item, to the hallec that Cato could afford to feed to his agricultural slaves. (10) Thus in the Diocletianic Price Edict first and second grade liquamen is listed at different prices. (11)

The ubiquitous use of garum, liquamen and their compounds in the recipes of Apicius suggest that they were an essential food requirement - at least for the Roman elite. (12) However, just as fish sauces are today widely consumed as a salt substitute in the countries of the Third World, so such sauces were consumed among the Roman lower classes to add flavour to their pulse and cereal based diets. Indeed at Pompeii evidence for fish sauces of various qualities has been found in the houses of both rich and poor, while shops existed solely for their sale. The papyri of Roman Egypt often refer to such products. (13) Of greater economic significance, fish sauce was a basic element in the diet of Roman soldiers; on a writing tablet from Vindolanda muria appears in a list of food requirements along with such staples as barley, wine and beer, while fish-sauce amphorae have also come to light from a number of military sites along the Rhine frontier. (14) The fact that Valentinian and Gratian banned the export of garum, along with that of wine and olive oil, outside the frontiers of the Roman Empire in the second half of the fourth century A.D. suggests that garum was considered to be of similar importance as a foodstuff to wine and oil. (15) In addition to their food value, garum and associated products were also used for a wide variety of medicinal purposes. (16)

It is possible now to gauge the importance of garum and other fish products archaeologically, by comparing the number of garum amphorae found in shipwrecks or at shore sites with that of wine and oil amphorae. The evidence only allows very rough orders of magnitude to be established, but it does seem to confirm the qualitative evidence already discussed and suggest that the demand for garum was at least commensurate with, if not greater than, the demand for wine and olive oil.

First, shipwrecks. Although marine archaeologists have tended to avoid rigorous quantification of cargoes of shipwrecks, a recent survey of 78 wrecks of Iberian origin found that for the period 50 B.C. to A.D. 250 the cargoes consisted of some 66 per cent. garum amphorae, compared to 26.5 per cent. olive oil amphorae and 7.5 per cent. metals. Once this evidence was broken down into fifty year periods, it was found that it was only in the period A.D. 150 - 250 that Baetican oil amphorae outnumbered fish-sauce amphorae. (17)

Secondly, Ostia. Two samples are available : the Augustan deposit at La Longarina and the material from the Baths of the Swimmer. First, La Longarina. This dump contains 168 Iberian amphorae, of which 104 contained garum or other fish products (61.9 per cent.), 58 contained wine (34.5 per cent.) and a mere six contained oil (3.6 per cent.). (18) Secondly, the Baths of the Swimmer. Panella has recently summarized the evidence for finds of amphorae in four periods. (19) The results of her findings are best expressed as a table (Table 5.1) :

TABLE 5.1 CONTENTS OF IBERIAN AMPHORAE FROM OSTIA : (A) LA LONGARINA (B) BATHS OF THE SWIMMER

(expressed as percentage figures)

PERIOD	FISH	WINE	OIL
A. LA LONGARINA			
1. Augustan	61.9	34.5	3.6

B. <u>BATHS OF THE SWIMMER</u>

1. Late Augustan	57	38	5
2. Flavian	59	32	9
3. Trajanic/Hadrianic	55	30	15
4. Antonine	70	6	24

NOTE : At Baths of the Swimmer Iberian amphorae constitute 50 per cent. of the total number of amphorae for Period 1, 21 per cent. for Periods 2,3 and 4.

The table confirms the testimony of the wrecks. *Garum* seems to have been shipped from the Iberian peninsula to Ostia in greater quantity than either wine or olive oil in the periods studied. *Garum* amphorae similarly predominate among the Iberian amphorae discovered at Pompeii and Castro Pretorio, Rome. (20)

However, recent quantification of amphora sherds from sites in Roman Britain presents a different picture with fish sauce amphorae thinly represented in the archaeological record. At Skeleton Green (Puckeridge, Hertforshire), for example, fish sauce amphorae make up only twelve per cent. (by weight) of all amphorae discovered, compared with thirty per cent. Dressel 20 Baetican oil amphorae. Gloucester is something of an exception with thirty per cent. fish sauce amphorae, but elsewhere the figure is seldom more than ten per cent. (21) Perhaps the reason for this is cultural : the native population of Roman Britain did not cultivate a taste for fish products, and thus the Roman military and administrative personnel constituted the main market for such products. This would at least explain the exceptionally high proportion of fish sauce amphorae at Gloucester, a Roman colony established for army veterans. (22)

iii. The Organization of Production

(a) Location and chronology of the industry

The literary evidence for fish processing in Lusitania is scant and not particularly revealing. Strabo recognized that it was practised west of the Straits of Gibraltar along the Algarve coast. (23) Since he was using Posidonius and Polybius as sources, his testimony suggests that the processing had begun at least by the early first century B.C.

Otherwise the main source of information is archaeological, the limitations of which need stressing at the outset. Many of the processing sites were discovered in the nineteenth century, with little attention devoted to stratigraphy or even to careful recording of finds. The evidence has been presented in the form of a catalogue raisonné (Appendix III), which gives a summary and full bibliography for each known site.

With the exception of the possible establishment at Peniche (C 23), no processing sites have yet been discovered north of the Tagus basin. However, more may still remain to be found, especially in those areas where the essential raw materials (fish and salt) were in plentiful supply : for example, in the estuaries of the Mondego and Vouga. Toponyms such as Marinha, Salinas (Aveiro), S. Felix da Marinha (Gaia) and Marinha das Ondas (Figueira da Foz) hint at the salt resources of this area. (24) Remains of fish processing sites have been discovered on the Galician coast north of Lusitania. (25) Overall, however, there seem to have been two main areas of garum production in Lusitania : the Algarve coast and the Sado estuary.

The configuration of the distribution of fish processing sites (Figure 5.2) provides a clue to their origins. For it corresponds very closely to areas of Phoenico-Punic influence, which has led to the assumption

FIGURE 5.2 FISH-PROCESSING SITES (CETARIAE) IN LUSITANIA

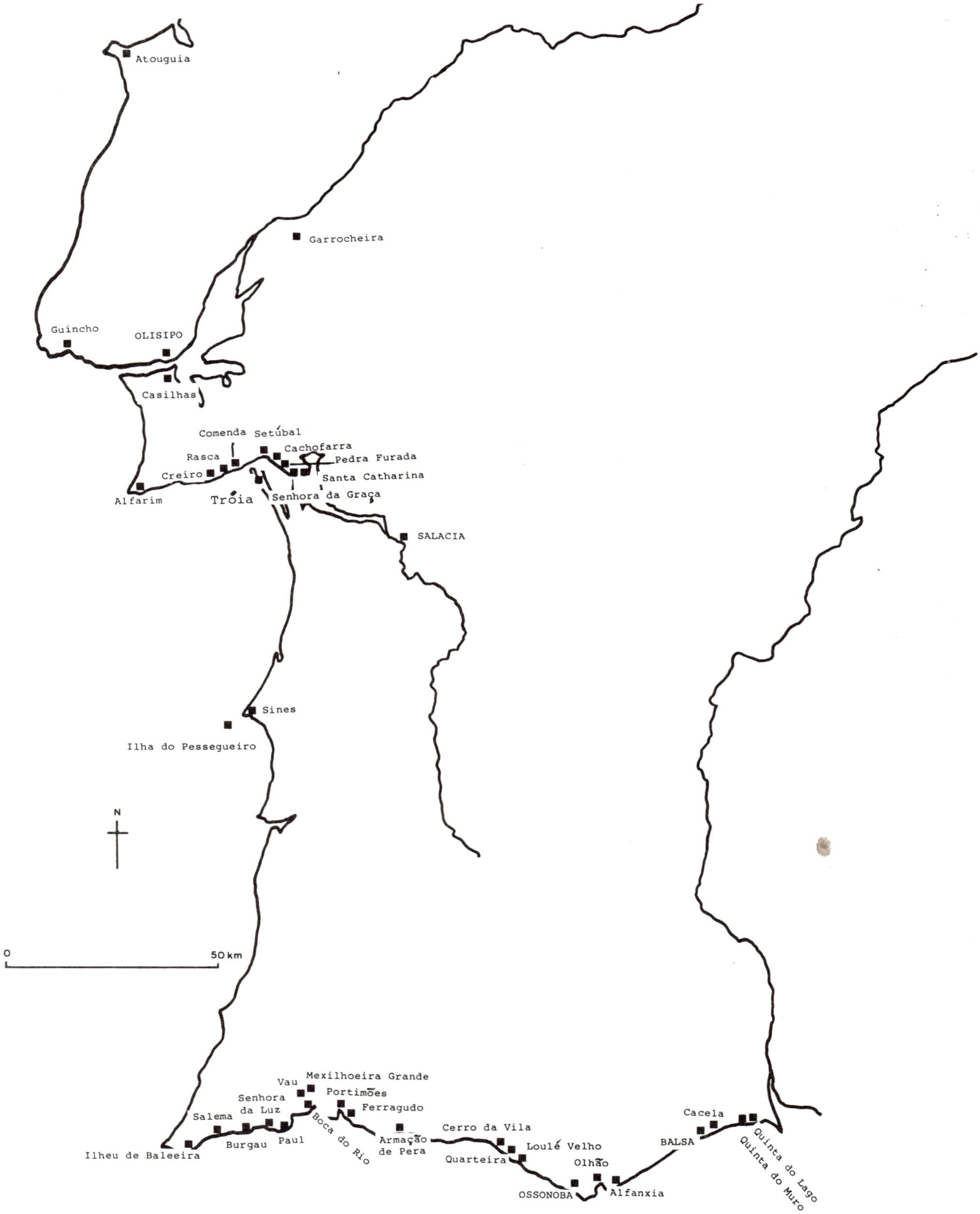

that the industry was initiated by these settlers. (26) Whereas in Sicily a Punic origin for fish processing sites has been confirmed by the discovery of Punic amphorae in the establishments themselves, as yet no similar confirmation can be made for the origins of those of Lusitania. (27) All the archaeological evidence points to purely Roman occupation.

But a number of factors combine to give some plausibility to the view that fish-processing developed under Phoenico-Punic influence. First, the earliest references to Iberian garum occur in the works of Athenian comic poets of the late fifth century B.C., a period when Phoenico-Punic settlers and merchants were already operative in southern Iberia. Recently remains of salted fish (especially sea-bream and tunny) have been found in early fifth century Punic amphorae, possibly produced in Spain, found in the excavations at Corinth. (28) Secondly, a Punic origin for the production of garum sociorum at New Carthage has convincingly been argued. (29) If the fish-processing industry had a Phoenico-Punic origin in other parts of the Iberian peninsula, the same probably holds true for Lusitania. Thirdly, the representation of tunny on coins with Turdetanian legends (for example from Gades, Abdera, Sexsi and Salacia) suggests that tunny fishing (and, a priori, processing) formed an important element in the local economies of these towns in the pre-imperial period. (30) Fourthly, if it can be demonstrated that purple dye was also fabricated in Lusitanian fish-processing establishments, this would strengthen the case for a Phoenico-Punic origin for fish salting, since the Phoenicians clearly played a predominant role in the production of purple dye in the Mediterranean world. (31)

But perhaps most suggestive is the growing body of evidence from Salacia (Alcácer do Sal) in the Sado estuary which points to significant Phoenico-Punic interest in the area. First, coins. The bronze coins of Salacia display almost identical types to those of Gades (Cadiz) : notably tunny fish and head of Hercules. Gades was the epicentre of Phoenico-Punic settlement in southern Spain. The suggestion is that Phoenico-Punic settlers from Gades migrated to Salacia and reproduced these coin-types. Moreover, the overstriking at Salacia of a coin from Punic Ebusus (Ibiza) illustrates further Phoenico-Punic links with Salacia. (32) Secondly, Phoenico-Punic amphorae dominate the finds from Republican levels on the acropolis of Salacia, again illustrating its links with the Phoenico-Punic world.

Only a few of the Lusitanian fish processing complexes have been excavated with sufficient thoroughness, and sufficiently recently, to allow an exact chronological sequence to be established. Therefore, the chronology of most sites is often based on the dating of a few fragments of pottery or coins. In the Algarve many of the sites have only been partly revealed because of the advance of the sea since Roman times. Here some sites occur as annexes to Roman villas. In these cases the chronology has to be based on that established for the annexed villa sites, although it does not necessarily follow that the industrial annex was in operation for the entire period of occupation of the villa. However, from the available evidence a general picture emerges that the industry was operating mainly from the first century A.D. until the start of the fifth century. To be more precise is difficult, since it is impossible to detect "phases of operation", periods of greater or lesser industrial activity.

On the basis of the archaeological evidence none of the processing sites were functioning during the Republic. Tróia (C 37) - the best documented site - was only occupied at the end of the first century B.C. Fragments of Arretine terra sigillata fineware discovered at Casilhas (C 25) and Quinta da Comenda (C 31), together with fragments of Dressel 1b Italian wine amphorae at Casilhas, suggest that these sites were also in operation by the end of the first century B.C. But elsewhere South Gaulish terra sigillata is the earliest form of fineware found, which suggests that these sites only came into operation in the first half of the first century A.D. (33)

Most of the sites can be shown to have still been occupied in the later fourth century A.D., with Tróia (C 37) continuing right through until the sixth century, although there may have been some lessening of activity from the fourth century onwards. The early Christian baptistry, with its stuccoed and frescoed walls, bears eloquent testimony to the continued flourishing of the site in the later Roman Empire.

The processing complexes of Setúbal provide some rather more precise data. The Praça do Bocage site (C 32a) went out of operation in the third century, after which the processing tanks were used as a rubbish dump. But the discovery of later Roman amphorae made of local Setúbal clay in third and fourth century levels suggests that other processing complexes were still operative in the town. This should alert us to the dangers of asserting that the "industry" was in decline, just because one site ceased operation; others, as yet undiscovered, may well have started to operate when the Praça do Bocage stopped. At the Travessa de Frei Gaspar site (C 32b) the tanks were found to have a false bottom, which has led to the hypothesis that there were two periods of operation : first, in the early Empire and, secondly, after a period of abandonment, in the fourth and fifth centuries. This example raises the salutary point that it does not necessarily follow that just because a site has yielded dated material from, for example, the first and the fourth century A.D., it was operational throughout that period. It might be tempting to use the data from these two sites at Setúbal to argue in favour of the widely held view of an economic "crisis" in the third century, but just because two sites went out of operation does not prove that the whole industry in the area was in a state of crisis. (34) On the island of Pesseguiero (C 40) the processing complexes were still operating until the mid fourth century.

(b) Raw Materials

One of the advantages of Roman fish sauces was that they could be made from a wide variety of fish, including demersal species, and even shell-fish, in addition to the pelagic species, on which most of the scholarly attention has centred. (35) It is now necessary to make some assessment of the availability of the essential raw materials : fish, shell-fish and salt.

Fishing now plays a central role in the Portuguese economy. Portugal is now the foremost exporter of fish conserves in modern Europe, while consumption of fish per head of population is only exceeded by the Norwegians and Japanese. (36) Recent statistics for fish caught in waters off the Portuguese coast serve to illustrate the variety of species available for processing, but also highlight some of the problems involved with fishing these waters. (see Table 5.3).

TABLE 5.3 FISH CAUGHT OFF PORTUGUESE COAST

(in tonnes) 1971-1980

YEAR		Dia-drome	Tunny	Cod	Mack-erel	Horse Mackerel	Sword-fish
1971	:	513	501	56961	20346	53796	4344
1972	:	345	206	49518	18614	59682	5062
1973	:	337	21	46704	21988	42236	6081
1974	:	270	42	49370	9437	48461	3757
1975	:	234	178	38668	7195	43911	4549
1976	:	243	262	40305	10216	49678	5346
1977	:	483	285	34813	11630	51766	6173
1978	:	308	503	30190	8539	32254	6770
1979	:	763	319	22710	6502	27389	5194
1980	:	658	4600	19480	4952	26549	5661

YEAR		Hake	Sar-dine	Others	Crust-aceans	Moll-uscs	Total
1971	:	31855	84523	72375	529	10902	348205
1972	:	27221	87597	69787	847	12205	342717
1973	:	44035	100950	85782	797	11353	375413
1974	:	21421	75489	64143	775	7228	297298
1975	:	18470	95486	55275	385	8356	285432
1976	:	27122	79246	54688	255	8656	286568
1977	:	15877	79823	67141	101	8424	290865
1978	:	19271	83600	52469	111	10520	255128
1979	:	14714	90523	56251	278	6475	244352
1980	:	28670	106324	48219	590	10518	256222

Source : Anuario Estatistico de Portugal

These figures illustrate one of the major problems of fishing in this area : that is, the great degree of inter-annual fluctuation in the quantities of individual species caught. Moreover, with less sophisticated fishing technology available to the Romans, *a fortiori* there would have been still wilder inter-annual variation in the size of the catch in Roman times. This would, therefore, have benefited those who could hold processed fish products in reserve against bad fishing seasons. The inter-annual fluctuation in the tunny catch is especially striking, but can be explained by the occasional, unforeseen changes in the pattern of migrations of this species. This unpredicability is to be expected still more in the Roman period and helps to explain the high prices that Roman producers could charge for tunny (and choice mackerel) products. (37) Some receipts from the tunny catch in the Algarve survive from 1490 and 1493 - 1496 : the figures vary from a low of 851,752 *reis* in 1493 to a high of 1,832,374 *reis* in 1496. (38) It is with these levels of fluctuation that the Romans would have had to have coped. (39)

The presence of tunny in Atlantic waters was noted by Oppian and Pliny. (40) The tunnies migrate from the Atlantic to the Mediterranean in late spring to spawn and then return to the Atlantic in July and August. So, as in Mediterranean, tunny catching off the coast of

Portugal is limited to two necessarily brief periods of the year. (41) The modern method of catching (in tunny traps) is not in essence any different from that outlined by Oppian, writing in the late second or early third century A.D. The trap nets are set in position (neither too near to, nor too far from, the shore); a look-out keeps watch from a watch-tower and on the appearance of the shoals informs the fishermen, who row out to sea and watch as the tunny pour blindly into the nets. (42) The use of tunny traps has become so integral to life in the Algarve that the Portuguese term for them (*armação*) has become enshrined in the toponym of a sea-shore settlement, Armação de Pera. Since this was the location of a Roman fish processing establishment (C 12), the continuity of economic activity from Roman times to the present is striking. In the late Medieval period also tunny traps were widely used in the Algarve, and Lagos was an important centre for the export of salted tunny. (43)

Other species of fish were also caught and processed. Mackerel was used for the fabrication of *garum sociorum*, while numerous *tituli picti* preserved on Iberian amphorae advertise that the fish-sauces contained therein were made from mackerel : for example, *gari scombri flos*, *gari scombri*, *liquaminis flos excellens scombri*, even *halec scombri*. (44) Traces of mackerel bones have also been found inside amphorae from the Port-Vendres II wreck. (45)

Sardines now form by far the greatest proportion of fish species caught off Portugal (see Table 5.3). Their great importance to the Portuguese economy is shown by the fact that there was a serious crisis in the Portuguese economy in 1948 and 1949, when no sardines at all were caught. (46) By analogy, it probably formed the basis of much of the Roman fish-sauce produced in the area. Traces of sardines have been discovered in an amphora from the Port-Vendres wreck. (47)

It is likely that such pelagic species formed the major proportion of fish caught, and processed, in the Roman period. For the fishing of demersal species, although it can be practised all year round (in contrast to the short fishing seasons for pelagic species), has a low level of productivity, so that it could only have been significant on an individual level, to supplement a family's diet, but could scarcely have provided a sufficient quantity of fish at any one time to make processing feasible. (48)

To sum up, Lusitanian waters could have provided a variety of fish species in sufficient quantities to allow processing in the Roman period, even though a consistent annual catch of any one particular species was even less possible than it is today. The inter-annual fluctuation of the tunny, and, to a lesser extent, mackerel caught gave a prestigious status and an added value to sauces prepared from them. The production of lesser quality fish-sauces would not have been hampered by such problems of supply, since any species of fish, and even the less choice parts of the fish, could have been used. The fact that so much of the fish caught (including parts that would otherwise have gone to waste, if the fish was consumed fresh) could have been processed in some way or other was a major economic advantage. Most production centres can, therefore, be expected to have produced all the various grades of fish-sauce, after the flesh of the fish had been salted. It would have been uneconomic to have concentrated on just one particular variety.

Because of its very seasonality, fishing is unlikely to have formed the major economic activity of a large number of the population throughout the year. Thus in modern times fishing is largely practised by individuals classed as agricultural workers on the Portuguese census. (49) So too in the sixteenth century the Duke of Medina Sidonia recruited fishermen from among the rural peasantry of Andalucia for the tunny catch. Furthermore, in modern Malaysia and Polynesia fishing is widely practised, but only as a supplementary activity to agriculture. (50) The rhythm of rural life in southern Portugal is also well-suited to a symbiosis of farming and fishing. For after the harvest in May, the peasants would not be occupied on the land until the vintage and olive gathering of the autumn and so would form an ideal source of labour for the summer's fishing season. Especially where fish-processing establishments were annexed to Roman villas, the estate owner would probably have had a source of labour in his agricultural workers, who could be used as fishermen and fish-processors.

The other major raw material needed was salt. The best salt for fish-processing is coarse-grain or "common" salt. This can be produced at lower temperatures than most other finer-grain salts, but it takes longer to crystallize. (51) For garum production two sextarii of salt were required for every modius of fish, which is a ratio of one part salt to eight parts fish. (52) Not surprisingly, most of the Lusitanian fish-sauce factories were located in the vicinity of salt resources, as also occurred in neighbouring Baetica. (53) The major modern salt producing areas of Portugal are the Algarve (especially around Faro and Lagos and in the Guadiana estuary), the Sado estuary (especially near Setúbal and Alcácer do Sal - a suggestive toponym) and the estuaries of the Tagus, Mondego and Vouga. (54) To date, it is only the estuaries of the Mondego and Vouga that have not yielded evidence for Roman fish-processing.

The connexion between salt extraction and fish-processing was noted by ancient authors, if sometimes only implicitly. Strabo, for example, reported that the Turdetanians had salt-mines and many streams of saline water, and then immediately remarked upon fish-processing in the area. (55) Pliny, by a chiastic and interlocking arrangement of his discussion of salt and garum, implicitly makes the connexion. (56) Manilius too mentions fish-sauces and then immediately turns his attention to salt-pans. (57)

For Lusitania there is no direct evidence of Roman salt production. Strabo perhaps gave a hint when he revealed that the Phoenicians regularly exchanged Iberian salt for tin from the Casseriterides. (58) The two major areas where Phoenicians established commercial entrepôts were Salacia (at the mouth of the Sado) and Santa Olaia, Figueira da Foz (at the mouth of the Mondego). Since these regions are today renowned for their salt resources, they are likely to have supplied the Phoenicians with part of their salt requirements. (59) Given the lack of salt mines in Portugal, the natural method of obtaining salt is, and always has been, by the evaporation of saline water along the sea-coast or in river estuaries. (60) Such a ubiquitous method is unlikely to have aroused the interest of Pliny or any other potential source.

However, the potential of the Lusitanian coast for salt production is clear from evidence available for the region from later periods. Salt was obtained on such a scale from the Sado estuary in the Middle Ages that from the thirteenth century a large surplus was exported to all parts of Europe. Although salt-pans were also worked in the estuaries of the Vouga, Mondego and Tagus and at various points along the Estremaduran coast and the Algarve, by far the highest levels of production were reached in the Sado estuary. On average twenty tonnes of salt were produced per year in each unit of production; that is five or six times the yield of the salt-pans of the Vouga estuary. The Portuguese crown held a monopoly of the industry and let out contracts to work the salt, for the most part, to monastic groups or towns. (61)

The salt was obtained by natural evaporation. The salt-pans were inundated in the winter/spring by the waters of the Sado, which were then left to evaporate until early August, when the first of three strainings took place, by raking off the crystallized salt. More salt water was then let in, and after twenty days there was a second straining. The remnants were then heated every two days to allow the final deposits of salt to be obtained. (62) The low level of technology involved in the method of obtaining salt makes it possible that the Romans employed a similar method in this area.

Production by this method was limited to the summer months, exactly when the resource would have been most in demand for fish-processing and when a labour force of unemployed peasants would have been available to carry out the work. (63)

(c) Revenue for Rome ?

Elsewhere in the Roman world salt-mines were the property of the Roman State and the right to work them was leased out to individual contractors (*conductores*) or, perhaps more commonly, to associations (*societates*). (64) In addition, that these contracts were sometimes leased by towns is suggested by the attestation of *salinatores civitatis Menapiorum* and *salinatores civitatis Morinum* from Gaul. (65) Overall, the system appears in many ways similar to that which governed the exploitation of the imperially-owned metal mines. Certainly the salt-pans of Byzantium were under the control of the kingdom of which it formed part, the local community and the local temple at various stages of their history. (66) But were the salt-pans of Lusitania controlled under a similar system ?

A text of the Digest includes salt-pans (*salinae*) among the resources that Roman landowners had to declare as part of their census returns. (67) Thus, if the salt resources were administered directly by the Roman State, revenue from the extra tax payable by those who owned such resources would have come under the general heading of *tributum soli* in the same way that the Roman State derived revenue from owners of vineyards and olive groves.

In more recent times the Portuguese crown controlled not only the salt resources (v. supra), but also fishing itself. In the sixteenth century the tunny nets were the property of the State and were leased to contractors and to associations. Thus the period of fishing could be rigorously controlled and a proportion of the catch could be easily be extracted as tax. (68) Similarly in Spain Philip II granted a monopoly of Andalucian tunny fishing to the Duke of Medina Sidonia. (69) More detailed information is available from the start of the twentieth century, when the modern states of Spain, Portugal and North Africa maintained a very close control over tunny fishing. In Tunisia, for example, a fiscal agent supervised the weighing of the catch, catalogued it and made the requisite deductions for the State. In southern Iberia the nets were still leased from the State, which also took a percentage of the catch as tax. The periods of fishing were thus closely regulated. (70)

Could the Roman provincial administration also have imposed such control on fishing in Lusitania ? Evidence from elsewhere in the ancient world suggest that governments could control the fishing of inland waterways and lakes in a similar way to that attested for southern Iberia in more recent times. One text of the Digest suggests that for lakes and inland waters the right to fish had to be purchased from the State; while another reveals that landowners had to declare as part of their census returns the value of any "fishing lakes" (lacus piscatorii) that they owned. (71) From Istria, an area of attested fish-processing, letters survive from governors of the province of Moesia to the local officials from the first century A.D. that dealt with the community's right to fish adjacent rivers and lakes. (72) Similarly, the right to fish the sacred lagoons of the Cayster was controlled from Ephesus, first through officials of the temple of Artemis and, later, by Roman publicani. (73)

All these pieces of evidence, however, concern fishing in inland lakes and rivers, and so they are not strictly applicable to sea-fishing. In Roman law the sea counted as public property (loca publica) and so was available for anybody to exploit, untaxed by the State. (74) The Roman provincial administration could not then have let contracts for the fishing of the open sea, but may still have derived some revenue from the activity as follows. An inscription survives from Troezen, datable to the second century B.C., which reveals that this Greek city derived revenues from the leasing out of tunny traps. (75) By analogy a similar process of leasing out tunny traps or other fishing equipment could have operated in Roman Lusitania.

Two possible modes of operation seem possible : either the provincial procurator could have let contracts for the control and leasing out of such equipment to individuals or associations (i.e. on a similar basis to their use of publicani for tax collection or to the system that prevailed in the area in more recent times); or the local communities could have been left to control the activity themselves as a way of boosting their local revenue, which the local magistrates could then have been used to help meet their overall fiscal liability to the Roman State. (76) On balance, the latter regime seems more feasible.

(d) Subsidiary Activities : Purple Dye ?

The discovery of molluscs of the _Murex brandaris_, _Murex trunculus_ and _Purpura haemastoma_ species at various Lusitanian fish-processing establishments raises the important question of the range of activities carried out at such sites. (77) In short, was the extraction and fabrication of purple dye from these molluscs a subsidiary branch of the fish-processing industry ?

Certainly these species of molluscs formed the major source of purple dye for the Romans, the importance of which is clear from the imperial measures taken to ensure its regular production. In the late Empire there came to be associations (_collegiae_ or _familiae_) of "purple gatherers" (_murileguli_), who were obliged to furnish a fixed amount of purple (the _canon conchyliorum_) for the State. Children became tied to the profession of their fathers, while imperial procurators supervised the entire production of various centres in both eastern and western Mediterranean. (78)

In addition to the archaeological discovery of these shells in fish-processing complexes, there are other factors which suggest that there was a symbiosis between the extraction of purple dye and fish-processing. The similarity of fabrication techniques is striking and significant, in that it would have allowed both purple dye and fish-sauces to be produced at the same establishment. This is brought out in Pliny's account of the fabrication of purple. First, the molluscs were collected in small pots containing bait, since they had to be caught alive to ensure that the small vein that contained the purple did not dry up. Then, the vein was extracted and macerated with salt for three days in tanks, after which water was added. The whole mixture was heated up at a uniform temperature for nine days, after which it was strained and the first trial dip of a fleece took place. (79)

This account points to three basic similarities : the need for large quantities of salt, the need for processing tanks and the need for the mixture to be heated at high temperatures. Pliny also reveals that the season for catching the requisite shell-fish took place in winter and spring. (80) This would have served to extend the period of operation of a fish-processing establishment, allowing for the purple processing to be completed just before the main fishing season began.

Another link between the two industries can be traced to the origins of both. The Phoenico-Punic settlers seem to have introduced both processes to the western Mediterranean. The Phoenico-Punic origin for fish-processing in southern Iberian has already been discussed. (81) The Phoenician interest in the purple industry needs little demonstration. The dominant position of their mother city, Tyre, is clear not only from the frequency of the literary references to Tyrian purple (so common that "Tyrian purple" soon became a generic term) or the representation of Murex and Purpura on the coinage of Tyre, but also from the enormous shell middens, consisting of waste material from the fabrication of purple, left standing at Tyre and Sidon. The Phoenician interest in purple has even been seen as an important determining factor in their expansion into the western Mediterranean, where further resources lay unexploited. (82)

Certainly the areas of Phoenician-Punic settlement in the western Mediterranean correspond closely to the location of purple and fish-processing industries. The island of Ebusus (modern Ibiza), for example, was settled by Phoenico-Punic immigrants; production of purple was so intense that in the later Roman Empire the island became the base of one of the procuratores baphiorum. (83) Similarly, the coast of Baetica was another area of Phoenico-Punic settlement, where fish-processing was a widespread economic activity and purple shell-fish were noted, for example, by Strabo. (84) Finally, in Mauretania fish-processing and purple manufacture were clearly practised in close proximity to one another. (85) The centrality of purple to the economy of the area is encapsulated in the toponym "Insulae Purpurariae", probably the islands of Mogador, originally a Phoenician trading post, where archaeological work has revealed the unmistakable traces of a fish-processing complex. Mogador provides the ideal example of a site of Phoenician origin which produced both fish-sauces and purple in its industrial installations. (86)

However, for Lusitania there is one important element in the jigsaw missing, namely large shell middens consisting entirely of Murex and Purpura species. Eight thousand molluscs were needed to produce just one gram of dye. (87) Therefore, if purple dye was being fabricated on a wide scale, large middens of shells left over from the processing are to be expected - as occurred at Tyre and Sidon. However, purple dye was certainly produced on the island of Leuke (off south-east Crete) in the Middle Minoan period, but no shell heaps survive. The absence of such middens can be explained on the grounds that the shells after the extraction of the vital vein could be reused in a number of ways. Recent discoveries at Berenice (Benghazi) vividly demonstrate that purple shells could be reused as a source for lime, or as pottery temper, or for construction fill. (88) An obvious re-use in Lusitanian fish-processing complexes would be for lime, necessary both for mortar, which was used to revet the salting tanks and the preparation area, and as temper for the firing of amphorae, which were made in the vicinity of the processing zones.

If purple dye was produced, this would considerably enhance the economic significance of the fish-processing complexes, since it would have allowed a more extensive period of operation during the year and would also have provided a luxury item, which would have fetched high prices. (89) But it would also have demanded further "industrial" activities in the vicinity. For purple dye was never transported as raw dye, but always in the form of dyed fabrics. This would thus presuppose textile production and fulling close to the dye works. Pliny's account of purple manufacture includes details concerning the dipping of fleeces. The literary sources single out the delicate fabrics of Salacia (Alcácer do Sal), an area noted for its fish-processing complexes. (90) As yet no archaeological confirmation of this is forthcoming, although there is some evidence for textile production at the villa at Montinho das Laranjeiras (Alcoutim) in the Algarve. (91)

An alternative explanation is that these molluscs were processed along with other fish into fish-sauces in the Roman period. (92) They were used, after all, as a foodstuff in Mesolithic times in the Tagus valley, and that remains their most common use today. (93) This hypothesis fits better with the heterogeneous composition of the few shell assemblages known from Lusitania (for example, that at Tróia) and also with the finds of such molluscs at Iron Age hill-fort sites (for example, at Rotura, Chibanes and Pedrão on the Setúbal peninsula). (94) Despite the low level of productivity involved with shell-fish in terms of labour input to calorie output (exacerbated by the high proportion (75 per cent.) of waste material to edible matter), many pre-industrial societies have used shellfish as a souce of food, not as a staple, but as a critical resource in times of shortage of other foodstuffs. (95) Thus, the processing of shell-fish probably formed a small, subsidiary activity of a fish-processing complex, practised just before the main fishing season. Since the Lusitanian coast is so well blessed with this resource (see Table 5.3), it was probably exploited in the Roman period, if only on a small scale and as a subsidiary to the processing of other kinds of fish.

(e) The Layout of a Fish-Processing Complex and the Organization of Production

With the exception of the vast complex at Tróia (C 37), most of the Lusitanian fish-processing sites are not sufficiently well-preserved to give a clear idea of their exact layout. Even with Tróia there are problems, since a plan of the entire site has never been published. The layout of a single unit of production at Tróia can be seen in Figure 5.5, and is discussed below. But for our best information on the layout of a complex as a whole we must turn to the well-excavated (and well-published) sites from north Africa.

The excavations at Lixus have revealed a large complex, consisting of at least ten individual units, arranged west to east along the Loukos estuary. These units consist of 147 tanks of differing dimensions, the total capacity of which has been estimated at some 1,013 cubic metres. The complex was instituted under Juba II, but did not start to flourish until the end of the first century B.C. Subsequently, there was considerable redesigning of the complex, which was utilized until the mid third century A.D. and, after a brief hiatus, until its final abandonment at the start of the fifth century. (96)

The complex at Cotta (illustrated in Figure 5.4) measured 56 metres long by 40 metres wide. The plan of the site clearly shows that it comprised some four main units, grouped around a central courtyard : preparation area, salting area, furnace and storerooms. Within these units, tanks of various dimensions are arranged around three sides of a central impluvium, where a cistern was located. All the tanks were constructed, as at Lixus, of "cocciopesto" (i.e. an impermeable mortar, made of lime and packed with small fragments of broken terracotta) and with all their corners rounded off, to prevent material conglomerating in any crevices. Both of these construction techniques are found generally in Lusitanian fish-processing complexes. A heating room with hypocaust has been excavated at Cotta (bottom right on the plan), as well as a small tower, which has been viewed as a

FIGURE 5.4 FISH-PROCESSING SITE AT COTTA, MAURETANIA

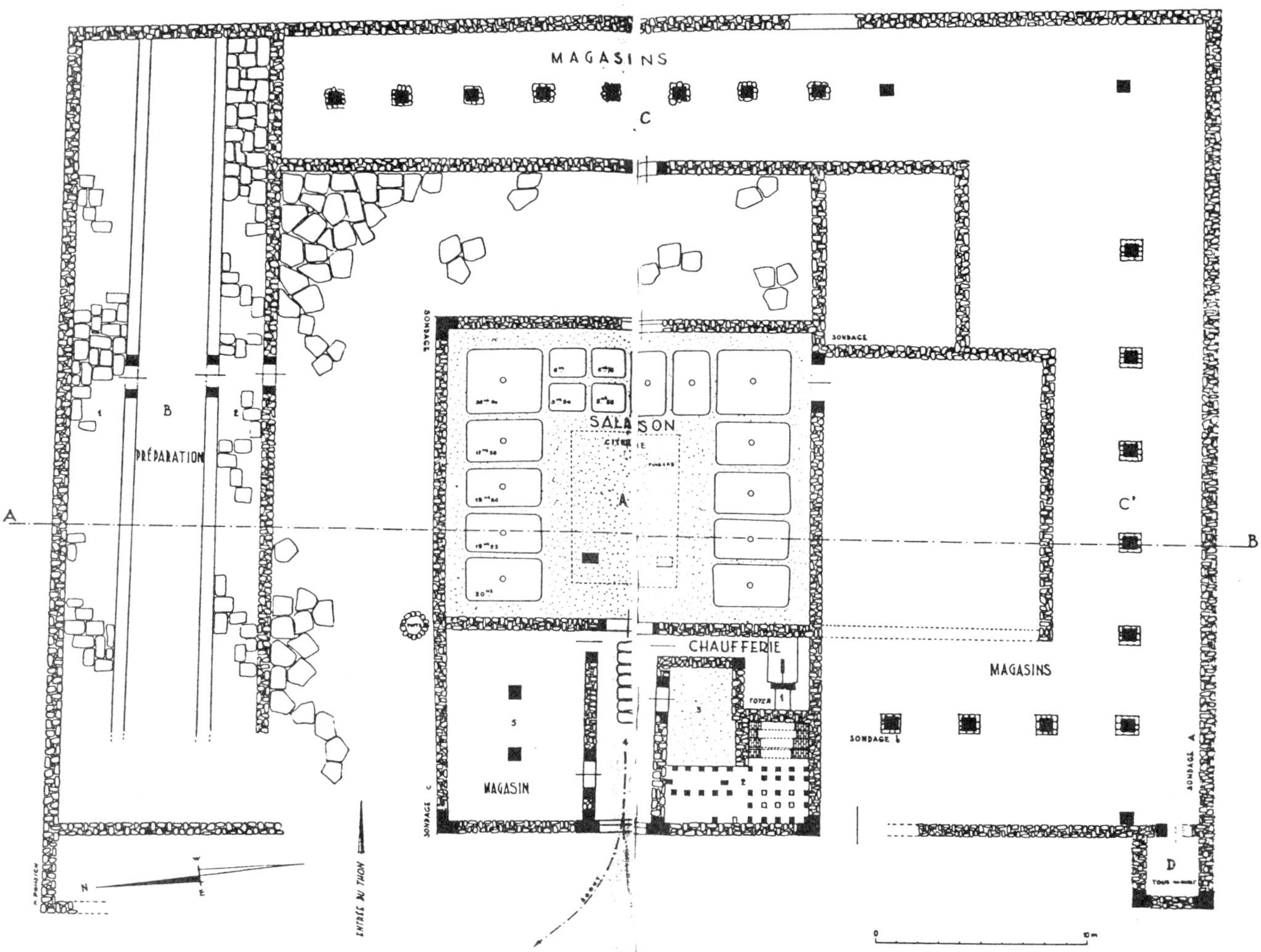

(Plan reproduced from M. Ponsich & M. Tarradell, Garum et industries antiques de salaisons dans la Méditerranée occidentale, Paris : Presses Universitaires de France by kind permission of the publishers)

watch-tower for the look-out to await the approach of the shoals, as outlined in Oppian's account of tunny fishing. (97) From material finds, the complex seems to have been operational from the end of the first century B.C. until the end of the third century A.D.

Thirdly, the complex at Tahadart consisted of six separate groups of tanks, arranged along the left bank of the river Tahadart. Only one group of tanks was fully excavated and the finds from here suggest that the tanks were in operation from the first century B.C. until the fourth century A.D. (98)

Even given the partial preservation of most Lusitanian processing complexes, it is, nevertheless, clear that their configuration and layout was very similar to that of the African examples here discussed. The plan of one unit of production at Tróia clearly demonstrates this (Figure 5.5). The fish would have been brought into the complex at the entrance (marked A); the preparation of the fish would have taken place in the rectangular court (marked B) : the floor of this area consisted completely of impermeable mortar/"cocciopesto", into which three holes had been sunk (marked C) to collect the left-overs from the washing and cutting up of the fish. Around this area were grouped the salting-tanks (marked D), also revetted with mortar/"cocciopesto". The bases of pillars (marked E) suggest that the salting-tanks were covered with a roof. This may have inclined inwards, as at Cotta, to direct water onto the preparation area to help keep it clean. But the presence of a roof would have precluded the natural evaporation of the saline/fish liquid, and suggests that a furnace would have been needed to allow the necessary maceration and evaporation.

It is clear from the plans of Tróia and Cotta that the salting-tanks were of differing sizes and capacities. This is best explained on the grounds that tanks of differing sizes were destined for different kinds, or strengths, of salted-fish or fish-sauce. So, the flesh of the fish, once cut into pieces, was salted and stacked in layers in the larger tanks, which contained brine, while liquid fish-sauces of varying strengths were fabricated from the scraps and insides of the fish mixed with brine in the smaller tanks. Possibly the smaller

FIGURE 5.5 FISH-PROCESSING SITE AT TRÓIA

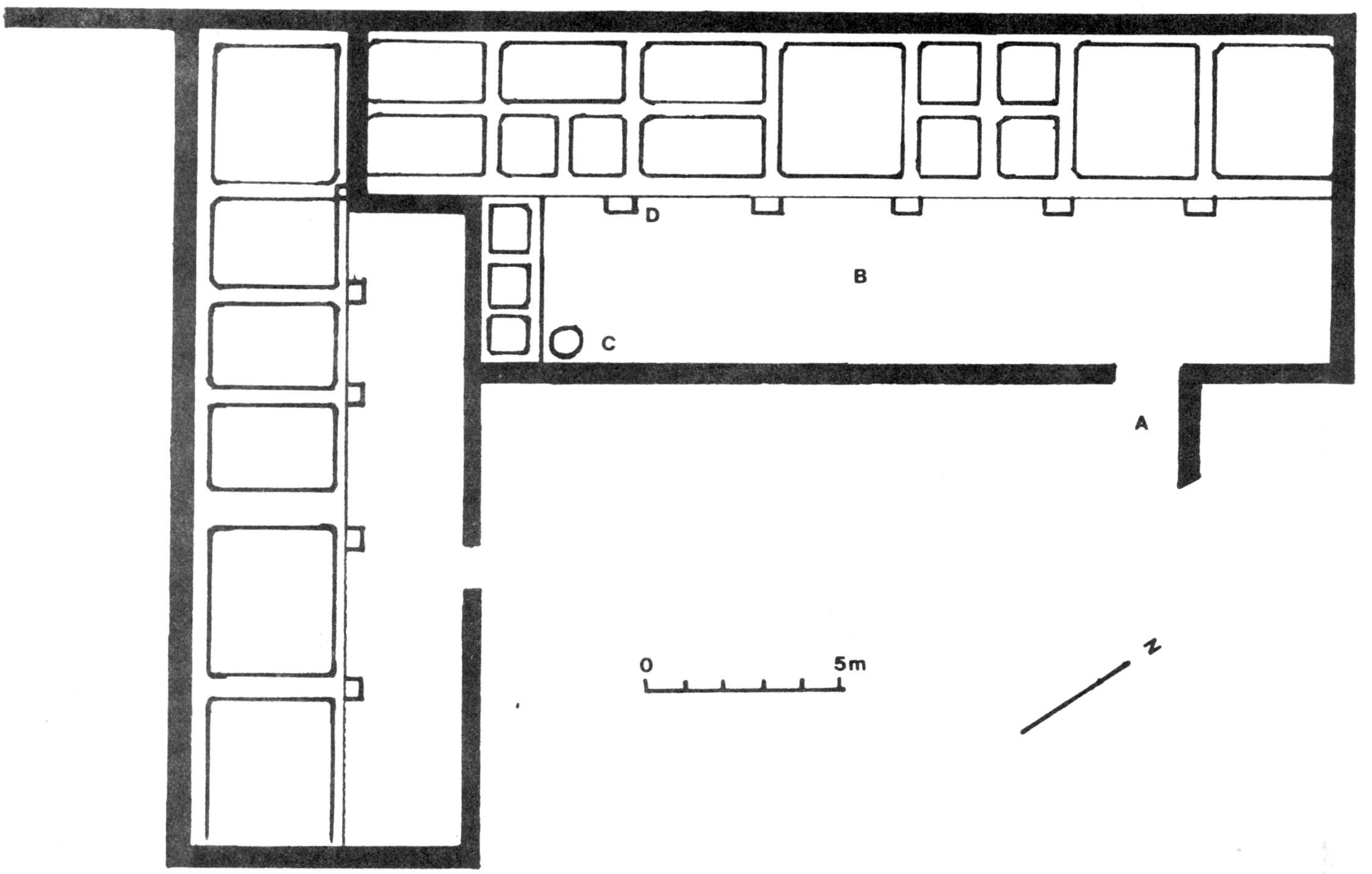

tanks contained the more concentrated (and, therefore, the more expensive) products.

The factors that led to such an arrangement of a fish-processing establishment are not easy to isolate, but if uncovered, could reveal much about the nature of the organization of production. (99) A passage of the Geoponika on the fabrication of fish-sauces is worth considering in this context in some detail, since it helps to explain some of the architectural features of the fish-processing complexes (see Appendix IV for an English translation). First, the fabrication process for liquamen is described as follows: the insides of fish are left to macerate with salt in a tank; garum is strained off first by sinking a basket into the tank and then liquamen in just the same way; the dregs that remain constitute hallec. (100) This would explain the variety in the size of the salting tanks : some were used for strained garum, others for liquamen, others for hallec. Secondly, the recipe for Bithynian garum is given : this involves the maceration of fish and salt in a tank for two to three months under a hot sun. (101) For processing on a large scale this period would have excessive and so it is not surprising to find that there was also a quicker process, whereby fish and oregano were added to a brine solution ("halme"), which was then heated up and strained. (102) In industrial complexes the latter method would no doubt have been practised, which explains the presence of furnaces not only at Cotta, but also at various Lusitanian processing complexes. Moreover, hypocausts have sometimes been reported in close proximity to these complexes. Rather than being associated with baths (the traditional explanation), it is probable that these hypocausts were needed to heat the furnaces used in the "industrial" processing of fish.

The need for such a furnace was an important factor in determining the architectural layout of a processing establishment. The expense of running and maintaining a kiln (especially in keeping it sufficiently supplied with fuel) led to a concentration of various units of production around a single furnace, as is clear, for example, at Lixus and Tróia. I have argued, with reference to the smelting of metal ores, that running a kiln was so expensive that it precluded the participation of all but the rich and caused smaller exploiters to rely on the use of, and to locate themselves in the vicinity of, the furnaces of the rich. (103) Although furnaces for fish-processing would not have required such enormous quantities of fuel (since the temperatures that needed to be reached were not so high), small operators would still have had to locate themselves in proximity to a furnace.

In other "industrial" contexts in the Roman world this process of nucleation has also been explained in similar terms. Thus, the availability of a nearby market, a good supply of fuel and ease of access to essential ancillary services have been seen as determining factors in the agglomeration of potters, for example, in one particular area. (104) The _terra sigillata_ workshops of Gaul provide a striking example of this kind of arrangement. At La Graufesenque eight separate units of production were grouped around a single kiln. (105) This type of layout suggests that the workshops were run by small-scale operators, perhaps united into associations (_societates_). Other evidence, such as the low quality of the tombstones and the generally modest levels of wealth revealed in the cemeteries near the pottery workshops of Lezoux and Rheinzabern, also points towards the same conclusion. (106)

By analogy, the layout of the large processsing complexes of not only Mauretania, but also Lusitania suggests that the work-force consisted of low-status individuals, sometimes working together in a co-operative enterprise. Just as miners and salt-workers formed themselves into associations, so workers in fish-processing may also have found it advantageous to band together. (107) This would, then, have allowed the participation of men of fairly low financial standing in the industry, operating small units of production. The very diversity of activity involved in fish-processing (fishing, salt extraction, processing) would have benefited from a co-operative approach.

The size of each unit of production remains a problem. For the _terra sigillata_ workshops of La Graufesenque, it has been recently suggested that each unit comprised five to ten workers. (108) In other locations of the industry the question is not so clear cut. But the two hundred kilns located in a relatively restricted area at Lezoux suggest that there were a large number of small units; at Rheinzabern the various installations of Comitialis seem to point to a similar organization of production. (109) But in contrast in the _terra sigillata_ workshops of Arretium (Arezzo) there seems to have been something of a coexistence of units of production of varying sizes and under a variety of modes of production. (110) It would thus be unwise to be too dogmatic about the size of a unit of production involved in Lusitanian fish-processing. A variety of modes of operation may have been involved in close proximity to one another at Tróia.

At Tróia the tanks for fish-processing extend for some four kilometres along a thin spur of land jutting into the Sado estuary. The exact number of tanks is hard to assess, but a nineteenth century traveller estimated (with no great precision) that the number exceeded two or three thousand. (111) The exact configuration of the entire complex cannot be ascertained until a detailed plan of the entire site has been prepared, but approximately fifty-two units of production have recently been estimated. (112) Given the greater diversity of activity involved in fish-processing than in fineware pottery manufacture, a slightly larger size for each unit of production seems likely for Tróia than at La Graufesenque. On the analogy of modern fish-processing in the seaboard communities of Malaysia, each unit of production may well have included the wives and dependents of the fishermen. (113)

There is, however, an alternative explanation for the clustering of small units of production around a central furnace. Nucleation may have come to pass not as a result of individual operators forming themselves into co-operative units, but at the instigation of a wealthy entrepreneur, so that it would enable him to employ as many free peasants, dependents or slaves in the fabrication of fish-sauces. In other words is this nucleation a sign of the investment of landed wealth in the industry and, if so, were the landowners indulging in what would now be termed "economies of scale" ?

The involvement of landed wealth in Lusitanian mining has already been discussed. (114) It was argued above that if a valuable resource was discovered on an estate, it seems a priori likely that a landowner would have been keen to exploit it by means of his own rural labour force at slack times of the agricultural year. So similarly where an estate was located near the coast and in proximity to salt resources, one would expect a Roman landowner to have become involved in fishing and fish-processing as a subsidiary enterprise to agriculture. The diversity of activity would have benefited from the sustained co-ordination and organization provided by such a regime. Furthermore, the initial cost of the wide variety of fishing equipment required (boats, nets, fishing-hooks) and the perennial need to repair and maintain such vital items demanded a substantial capital outlay. In addition, considerable quantities of raw materials were needed, to make the initial investment in such equipment worthwhile. All this suggests that a landowner would be much better placed to control such a diversified operation than would groups of individuals of a lower social status and economic position. (115)

The direct involvement of landowners in the industry can best be demonstrated in the Algarve. Where villas occur on the coast or in river estuaries, archaeological remains of fish-processing complexes have often been found as industrial annexes to the residential centre of the estate. A good example of this is the villa at Boca do Rio, Budens (PS 34), but the connexion can also be made at other sites (see Figure 5.6). It is unlikely that fish-processing formed the major economic activity on these estates, but rather an enterprise subsidiary to agriculture. Here the landowner would have had an obvious source of labour in his agricultural peasants at slack period of the agricultural year. (116)

TABLE 5.6 FISH-PROCESSING COMPLEXES LOCATED IN PROXIMITY TO ROMAN VILLAS

	Site *	Villa	Ref.+
1.	C 06	Budens, Boca do Rio	PS 34
2.	C 08	Abicada, Mexilhoeira Grande	PS 38
3.	C 09	Vale de Arrancada	PS 40
4.	C 10	Castelo de Arade, Ferragudo	ARA, I, 135-6
5.	C 12	Cerro da Vila, Quarteira	PS 41
6.	C 16	Quinta do Marim, Olhão	PS 39
7.	C 19	S. Estevão, Paul (Tavira)	PS 42
8.	C 20	Quinta da Monte Rosa, Cacela	PS 35

NOTE : * : refers Catalogue of Fish-Processing Sites (Appendix III)

+ : refers to J.-G. Gorges, Les villas hispano-romaines, Paris, 1979 : sv. Catalogue.

Grenier held that peasants, whose fundamental activity it was to work the land, often performed other tasks during the agricultural off-season, of which garum production was one. His case is considerably strengthened if one accepts his interpretation of a titulus pictus on a Spanish amphora from Augst, which he sees as providing evidence for an estate manager (vilicus) in Spain overseeing the production of fish-sauce. (117) Pottery manufacture is another possible activity for idle peasants. A text of the Digest suggests that potters were not employed throughout the year in pottery manufacture, but would also have worked in the fields. (118) This was perhaps the way in which production of fineware pottery was organized at La Graufesenque. (119) A papyrus from Oxyrhynchus in Roman Egypt shows that pottery manufacture took place there in the winter months, when there was little to be done on the land. (120)

Few traces of permanent habitation have been discovered in the vicinity of fish-processing establishments, from which it has been argued that the work-force for both fishing and fish-processing lived in the nearest urban centre or, if the processing establishments were located far from an urban centre, in wooden huts temporarily erected just for the processing season. (121) The lack of permanent habitation in the vicinity of processing establishments could alternatively be explained on the grounds that agricultural workers provided the labour for the processing. Thus, they would have continued to live in their normal dwellings on the land, dwellings so modest that only very occasionally have they left their trace in the archaeological record.

In the other major area of Lusitanian fish-processing (the Sado/Tagus basin) and especially at Tróia, the largest fish-processing complex in Lusitania, and second largest in the Iberian peninsula after Bolonia (Belo), the involvement of landed wealth is less easy to detect. From here there is no evidence for such clear-cut connexions between villa owners and fish-processing as in the Algarve. This is not particularly surprising, since few Roman villas have been located in this area of general unpromising agricultural conditions. In contrast to the Algarve, the industrial complexes seem to have been located in urban or at least "semi-urban" contexts : thus the sites at Setúbal (probably Roman Caetobriga), Tróia, Lisbon (Roman Olisipo) and Alcácer do Sal (Salacia).

However, in the immediate vicinity of the processing complexes at Tróia, houses of some elegance have been discovered, their mosaics and wall-paintings suggesting a relatively elevated financial standing for their owners. (122) The very location of these houses makes it probable that their owners derived at least some of their wealth from the local industry. The presence of such houses, a large bath house, two burial chambers (columbaria) and a later baptistery suggests that the site was occupied for more than just that brief part of the year which constituted the fishing season. These features, in addition to the sheer size of the complex as a whole, point to permanent occupation, and perhaps also continuous operation, throughout the year. Subsidiary activities (for example, the preparation of shell-fish into sauces or, possibly, purple dye) would have ensured that there was a full level of production for most of the year. In short, the size and nature of the settlement at Troia suggests that the industry was significant enough not only to stimulate the development of a semi-urban community (an industrial vicus ?), but also to attract a sizeable local population to become involved in the processing procedures or in the provision of supplies or ancillary services for the industrial community.

The current prevalence of sandy, saline soils in the hinterland of Tróia does not encourage cereal cultivation. If it is assumed that a broadly similar ecological pattern prevailed in Roman times, then considerable attention would have had to have been paid to the importation of vital foodstuffs for the local community. Since fish products from Tróia seem to have found their way into the cargoes of Baetican and, later, North African shipowners (as shall be seen in Chapter Six), they could have been exchanged for Baetican and North African grain, wine and olive-oil required. The presence in Troia of Baetican and North African oil amphorae gives a hint of this exchange. (123)

Finally, some hints as to the social make-up of the community at Tróia can be gleaned from the (admittedly small) epigraphic record. First, the nature of the inscriptions themselves suggest that we are dealing with individuals of a fairly modest level of wealth. For the inscriptions consist either of small stone altars of little elegance or of small plaques to be set into such altars. Secondly, the onomastics are revealing. The evidence is best tabulated (Table 5.7).

TABLE 5.7 NAMES ATTESTED ON INSCRIPTIONS FROM TROIA

1.	L. Iulius L. et T. lib. Valens	freedman	CIL II 41 IRCP 211
2.	Copirus (brother of 1)	slave	CIL II 41 IRCP 211
3.	G. l. Fabianus	freedman	CIL II 42 IRCP 215
4.	Paccia Lepidina	freedwoman ?	IRCP 219
5.	Q. l. Felix (grandson of 4)	freedman	IRCP 219
6.	L. Attius Helvianus	freedman ?	IRCP 208
7.	C. Servilius Claranus	freedman ?	IRCP 220
8.	L. Vibius Callistratus	freedman ?	CIL II 44 IRCP 221
9.	Marius Antiochus	freedman ?	CIL II 43 IRCP 217
10.	Aponius Seronus	freedman ?	IRCP 214
11.	Licinia (wife of 10)	?	IRCP 214
12.	Hypnus	slave ?	EE VIII 5 IRCP 210
13.	Galla (wife of 12)	slave ?	EE VIII 5 IRCP 210
14.	Licinia Galla ?	freedwoman ?	IRCP 213
15.	Galla ?	?	IRCP 224
16.	Liberius (or Liberinus) (dies at 5)	?	IRCP 212

17.	Marciana (dies at 3)	daughter of freedman ?	IRCP 216
18.	Mersina (or Mersinta) (dies at 9)	slave ?	IRCP 218
19.	Emerita ??	?	IRCP 209
20.	Maternus/Materna	son/daughter of 19	IRCP 209
21.	Parthenopaeus ?	slave	IRCP 222a

NOTE : Some of these inscriptions have only recently been published. I am most grateful to Dr. d'Encarnação of the University of Coimbra and to the Centre Pierre Paris, Bordeaux, for information on them prior to publication.

The most striking fact of this list is the high proportion of freedmen and freedwomen. Only three are self-confessedly freedmen, and thus of servile origin, (nos. 1, 3, 5), while two others are related to these freedmen, and hence also of servile origin : Paccia Lepidina (no. 4) is the grandmother of Q. l. Felix (no. 5) and Copirus (no. 2), a slave, is the brother of L. Iulius L. et T. l. Valens (no. 1). I attribute probable libertine status to L. Attius Helvianus (no. 6) and Marciana (no. 17) on the grounds that names ending in "-ianus" are often borne by freedmen, to C. Servilius Claranus (no. 7), L. Vibius Callistratus (no. 8), Marius Antiochus (no. 9) and Aponius Seronus (no. 10) on the grounds that "a Greek or Oriental name borne by a resident of a western city points to descent either from a free immigrant from the east Mediterranean region or a slave". (124)

If this is right, the presence of freedmen is consistent with what is a common feature of other urban, industrial concerns in the western Roman Empire, namely that libertine workers often rose to positions of authority and responsibility. (125) The presence of women and children raises the possibility that they may been been involved in the industrial activities, as well as men, as is common in modern Malaysia. (126) Alternatively, even if they were not themselves involved in the processing, it would at least confirm that the fish-processing site at Troia was occupied on a permanent basis.

Given the presence of these freedmen and freedwomen, the question remains as to their exact role in the operations at Tróia. It is traditionally held that freedmen in industry acted as commercial agents for wealthy patrons, who manipulated and controlled their actions, and took a large slice of the profits. (127) But there was a wide range of possible relationships between freedmen and their patrons; sometimes freedmen were able to act with a considerable degree of independence in commerce or in manufacturing ventures, even when they were acting as business agents or managers (procuratores, institores or actores) for their former patrons. (128) The evidence from Tróia does not allow us to assess whether those freedmen attested were "independent" (i.e. they had invested their peculium into a small industrial enterprise of their own) or whether they were acting on behalf of a wealthy landowner, who controlled a larger-scale operation.

The presence of these freedmen does not exclude other modes of production. We should rather imagine there to have been a whole range of different forms : ranging from small-scale operations run by low status, freeborn men, sometimes uniting into associations, right through to large units of production, controlled by wealthy local landowners often through estate managers. Slave, libertine and free workers were probably all involved. (129) Some freedmen operated independently, others under contractural relationships to former masters. That there were family units of freedmen is suggested by the epigraphic evidence tabulated in Table 5.7. That the complex at Tróia was integrated within a monetized economy is suggested not only by scattered coins found on the site, but by the coin hoards dicovered near the industrial workshops. (130)

The scale of production at Tróia led to a much more diversified and complex organization of production, involving a considerable degree of division of labour, than found in any establishment annexed to a landed estate. More opportunities for slaves, freedmen or free workers would have presented themselves in the urban (or semi-urban) environment at Tróia, Caetobriga, Olisipo and Salacia than on the rural estates of the Algarve.

(f) Conclusion

The discussion of the organization of production has thrown up some possible configurations for the size and nature of the units of production. On the whole, the units seem to have been small, as in other industrial concerns in the Roman West, but they were sometimes nucleated either for the mutual benefit of small operators or for the convenience of wealthy individuals who controlled such units of production. It is difficult to be precise over the exact number of workers involved in each unit, especially since many ancillary services would have had to have been provided, as in mining communities, to keep such a large workforce supplied. Much of the discussion has centred on Troia : the physical extent and layout of the site, as well as its long period of occupation and operation, give some indication of its considerable importance to the local economy.

However, not all Lusitanian fish-processing complexes were organized on such a scale and many (especially in the Algarve) were clearly industrial annexes of landed estates. In these situations fish-processing would have complemented other, more important, agricultural activities on the estate and may have been organized along similar lines to the production of wine and olive-oil. (131) The scale of production was obviously smaller than at complexes such as Tróia, Bolonia or Cotta and was probably carried out only in the summer months after the harvest. Fish processing could also have been carried out on imperial estates, if they were located in proximity to the necessary raw materials. As yet there is no evidence for this from Lusitania, but an amphora from the Saint Gervais wreck provides a clear-cut case from Baetica. (132)

The difference between the organization of production at the processing sites of the Algarve and those at Tróia may well reflect the fact that the former were essentially rural, estate-linked, while the latter formed part of a more developed urban economy. The difference in scale between the two areas means that it is perhaps only legitimate to talk of a fish-processing "industry" in the case of the Sado estuary, where a heavy concentration of units of production grew up in a relatively restricted area. It would interesting to investigate the Tagus estuary in more detail : for one suspects that a heavier density of fish-processing sites may well have been located here in Roman times.

Unfortunately the urban sprawl of Lisbon makes the chances of success of such an investigation limited.

Finally, the presence of two major ports (Olisipo and and Salacia in the Tagus and Sado estuaries respectively) was a significant factor in the development of the industry in these two areas, since any surplus produce could easily be disposed of. Surplus from the Algarve sites could be shipped from two smaller ports : Portus Hannibalis (Portimão) or Ossonoba (Faro).

NOTES : CHAPTER FIVE

** It should be noted that in this and subsequent chapters I use the term "garum" loosely, to refer to all kinds of Roman fish-sauce, rather than specifically to refer to the top-grade product. It serves as a convenient short-hand for "garum and other types of fish-sauce".

1. See, for example, Strabo 3.2.7 (145); Oppian, Hal. 3.620-622; Polybius 34.8.6-7 ap. Athen. 8.302e; for fish in the Tagus : Str. 3.3.1 (153); Plin. NH 9.4.9-12; see in general A. Schulten, Iberische Landeskunde. Geographie des antiken Spanien. II, Strasbourg, 1957, 572-7.

2. Baesuris (= Aesuri on legends) : A.M. de Guadan, La moneda ibérica : catalogo de numismatica ibérica e ibero-romana, Madrid, 1980, 221; A. Vives y Escudero, La moneda hispánica, III, Madrid, 1924, 81; Ossonoba : Guadan, 238, nos. 917-918; Vives, 114; Salacia : Guadan, 35-36, nos. 125-126; Vives, 24-27, nos. 1-4; Myrtilis : Guadan, 230, nos. 895-898; Vives, 90-91. Lead tesserae from Ossonoba reproduce the same type as the coins : M.L.E. da Veiga Affonso dos Santos, Arqueologia romana do Algarve, I, Lisbon, 1971, 203-204, 261. Similar tesserae from Balsa : ibid., 260-261.

3. O. da Veiga Ferreira, 'Algumas notas acerca da pesca na antiguedade', AP, iii, 2, 1968, 113-133; M. Ribeiro, 'Anzöis de Tróia : subsidios para o estudo da pesca no periodo lusitano-romano', AP, iii, 4, 1970, 221-236.

4. J.R. Coull, The Fisheries of Europe : an economic geography, London, 1972, 176.

5. A.H. de Oliveira Marques, Daily Life in Portugal in the late Middle Ages (tr. S.S. Wyatt), Wisconsin, 1971, 22; R. Firth, Malay Fishermen : their Peasant Economy, London, 1946. In general see I.M. Mackie,

R. Hardy & G. Hobbs, 'Fermented fish products', F.A.O. Fisheries Report, No. 100, Rome, 1971, 1-54.

6. R. Firth, op.cit. (n. 5), 10.

7. On salsamenta see Varro RR 3.17.7; Col. 8.17.12; Plin. NH 32.2.9; dealers in salsamenta (salsarii or salsamentarii) are attested, for example, at Rome : CIL VI 9677, 9873, 9676.

8. See RE VII 841-849 (garum), XVI 661-662 (muria), I 1584 (allec). On lymphatum see D. Manacorda, 'Anfore spagnole a Pompei' in L'instrumentum domesticum di Ercolano e Pompei nella prima eta imperiale (= Quaderni di cultura materiale, I), Rome, 1977, 121-133, esp. 127. On Roman fish-sauces in general see R.I. Curtis, Production and Commerce of Fish Sauce in the Western Roman Empire (unpubl. Ph.D thesis), University of Maryland, 1978; P. Grimal & T. Monod, 'Sur la véritable nature du garum', REA, 54, 1952, 27-38; C. Jardin, 'Garum et sauces de poisson de l'antiquité', RStLig, 27, 1961, 70-96; J. André, L'alimentation et la cuisine à Rome (2nd ed.), Paris, 1981, 109-113, 195-198; R.I. Curtis, 'In defense of garum', CJ, 78, 1983, 232-40.

9. Oleogarum : Apic. 32, 73; oenogarum : Mart. 7.28.3; oxigarum : Athen. Deip. 9. 366c; hydrogarum : SHA, Sev.Alex. 40.6; laccatum : D. Manacorda, art.cit. (n. 7), 127, citing Apul., De medic.herb. 4.12.

10. R. Étienne, 'A propos du "garum sociorum"', Latomus, 29, 1970, 297-313; J.R. Garcia del Toro, '"Garum sociorum" : la industria de salazones de pescado en la edad antigua en Cartagena', Anales de la Universidad de Murcia, 36, 1978, 27-57; Cato, RR 58.

11. 3.6 (liquamen primum); 3.7 (liquamen secundum).

12. Apicius, De Re Coq. (passim). The provincial elite would also form a market : for the example of Ausonius of Bordeaux see R. Étienne, 'Ausone et l'Espagne' in Mélanges d'archéologie, d'épigraphie et d'histoire offerts à Jerome Carcopino, Paris, 1966, 319-332, esp. 324-325.

13. For example, nuoc-mam in Vietnam : P. Grimal & T. Monod, art.cit. (n. 8), 31-34; C. Jardin, art. cit. (n. 8), 87-89; belachan in Malaysia : R. Firth, op.cit. (n. 5), 10; soumbala in tropical Africa : P. Grimal & T. Monod, art.cit. (n. 8), 38. For Pompeii : R.I. Curtis, 'The salted fish industry of Pompeii', Archaeology, 37.6, Nov./Dec. 1984, 58-9, 74-5; for Egyptian papyri : R.I. Curtis, art. cit. (n. 8), 236-7.

14. A.K. Bowman & J.D. Thomas, Vindolanda : the Latin Writing Tablets (= Britannia Monograph, 4), London, 1983, 83-93, tablet 4, esp. line 33. In general see R.W. Davies, 'The Roman military diet', Britannia, 2, 1971, 122-142, esp. 131.

15. CJ 4.41.4.

16. C. Jardin, art.cit. (n. 7), 92-96; Plin. NH 31.44.96-97.

17. R. Pascual Guasch, 'La evolución de las exportaciones béticas durante el Imperio' in J.M. Blazquez (ed.), Producción y comercio del aceite en la antiguedad (I Congreso Internacional, Madrid, 1980), Madrid, 1981, 233-242, esp. 237-240.

18. A. Hesnard, 'Un dépôt augustéen d'amphores à la Longarina, Ostie' in J.H. d'Arms & E.C. Kopff (ed.), The Seaborne Commerce of Ancient Rome : studies in archaeology and history (= MAAR, 36), Rome, 1980, 141-156, esp. 149.

19. C. Panella, 'La distribuzione e i mercati' in A. Giardina & A. Schiavone (ed.), Merci, Mercati e Scambi nel Mediterraneo (= Societa romana e produzione schiavistica, II), Rome/Bari, 1981, 55-80, esp. Tav. XV. See also C. Panella, 'Annotazioni in margine alle stratigrafia della Terme ostiensi del Nuotatore' in Recherches sur les amphores romaines (= Collection de l'école française de Rome, 10), Rome/Paris, 1972, 69-106.

20. D. Manacorda, art.cit. (n. 8), esp. 121 (Pompeii), 123 (Castro Pretorio, citing E. Dressel, 'Di un grande deposito di anfore rinvenuto nel nuovo quartiere del Castro Pretorio', *BCAR*, 7, 1879, 36-112, 143-196).

21. D.F. Williams & D.P.S. Peacock, 'The importation of olive oil into Iron Age and Roman Britain' in *Producción y comercio del aceite en la antiguedad II*, Madrid, 1983, 263-80, esp. 272. For full details of Skeleton Green see C. Partridge, *Skeleton Green* (Britannia Monograph, 2), London, 1981.

22. For Spanish soldiers in Britain see J.M. Roldan Hervas, *Hispania y el ejercito romano*, Salamanca, 1974, 134-45. For Lusitanian amphorae from military camps at Vindolanda and Beckfoot, Cumberland see D.R. Wilson, R.P. Wright & M.W.C. Hassall, 'Roman Britain in 1973', *Britannia*, 5, 1974, 397-480, esp. 467, note 41.

23. 3.2.6 (144).

24. See A. do Amorim Girão, *Atlas de Portugal* (2nd ed.), Coimbra, 1958, Map 28 : "Industrias extractivas"; on salt and fishing at Aveiro at the mouth of the Vouga in more recent times see L. de Magalhães, 'Os barcos da ria de Aveiro', *Portugalia*, 2.1, 1905-1908, 49-62; for toponyms H. de Barros Bernardo, 'Marinhas ignoradas da Estremadura - as salinas de Peniche', *Ethnos*, 5, 1966, 121-67, esp. 123.

25. A. Tranoy, *La Galice romaine*, Paris, 1981, 244.

26. M. Ponsich & M. Tarradell, *Garum et industries antiques de salaison dans la Méditerranée occidentale*, Paris, 1965, 89-90 & fig. 58 (p. 114); A. Garcia y Bellido, *Fenicios y Cartagineses en Occidente*, Madrid, 1942, 82-93.

27. See G. Purpura, 'Pesca e stabilimenti antiche per la lavorazione del pesce in Sicilia. I. S. Vito (Trapani), Cala Minnola (Levanzo)', *Sicilia Archeologica*, 48, 1982, 45-60, esp. 53, 57.

28. Eupolis (446-411 B.C.) : T. Kock (ed.), Comicorum Atticorum Fragmenta, Leipzig (3 vols.), 1880, 1884, 1888, I, 310, fr. 186; Antiphanes (fl. c. 390 B.C.) : ibid., II, 43, fr. 77; Nicostratus, son of Aristophanes (fl. 380 B.C.) : ibid., II, 220, fr. 4; for Corinth see C. Kaufman Williams II, 'Corinth 1978 : Forum Southwest', Hesperia, 48.2, 1979, 105-44, esp. 117-8 & Pl. 46; C.G. Koehler, 'Corinthian developments in the study of trade in the fifth century', Hesperia, 50.4, 1981, 449-58, esp. 450.

29. R. Étienne, art. cit. (n. 10), 311-313.

30. Gades : A. Vives y Escudero, op.cit. (n. 2), III, 8-12; Abdera : ibid., III, 16-19; Sexsi : ibid., III, 19-22; Salacia : ibid., III, 24-27.

31. See below, p. 118-21.

32. For coin types see note 30. For overstriking : E. Collantes Vidal, 'Reacuñaciones en la moneda ibérica', Ampurias, 31-32, 1969-70, 255-257, esp. 256.

33. For detailed references to this material see Catalogue (Appendix III) under individual sites.

34. For the theory see, for example, J.M. Blazquez, Estructura económica y social de Hispania durante la anarquia militar y el Bajo Imperio, Madrid, 1964.

35. For fish-sauce made from tunny : Geop. 20.46.6; mackerel : Plin. NH 31.43.94; Mart. 3.50.4; Str. 3.4.6 (159); Geop. 20.46.3; mullet : Plin. NH 31.44.95; Sen. Quaest. Nat. 3.17; sprat : Geop. 20.46.1; oysters and shell-fish : Plin. NH 31.44.95.

36. E.N. Shanahan, Western and Central Europe : a regional geography, London, 1964, 421.

37. For price : Pliny NH 31.43.94 : HS 1,000 for 2 congii (= c. 6.5 litres) of garum sociorum. In

general Sen. Ep. 95.25.

38. J.A. Romero Magalhães, Para o estudo do Algarve economico durante o século XVI, Lisbon, 1970, 149 (cf. 1,394,420 rs. in 1490, 1,002,589 rs. in 1494, 1273,707 rs. in 1495).

39. On the unpredictability of the resource base see T.W. Gallant, A Fisherman's Tale : an analysis of the potential productivity of fishing in the ancient world (= Miscellanea Graeca, 7), Ghent, 1985.

40. Opp. Hal. 3. 620-622; Plin. NH 31.43.94.

41. J. Bourge, 'Étude sur les migrations du "scomber thynnus" (Thon commun) dans le bassin occidental de la Méditerranée', Révue Tunisienne, 1915, 85-97.

42. Opp. Hal. 3. 633-648

43. F. Braudel, The Structures of Everyday Life : the limits of the possible (= Civilization and Capitalism 15th - 18th Century, I), (tr. S. Reynolds), London, 1981, 215.

44. All attested on amphorae from Pompeii : D. Manacorda, art cit. (n. 8), 126.

45. D. Colls, R. Étienne, R. Lequément, B. Liou, F. Mayet, L'épave Port-Vendres II et le commerce de la Bétique à l'époque de Claude (= Archaeonautica I), Paris, 1977, 40-42.

46. G.W. Hoffman, A Geography of Europe, including the Asiatic U.S.S.R. (2nd ed.), London, 1961, 535.

47. Y. Chevalier & C. Santamaria, 'L'épave de l'anse Gerbal à Port-Vendres', Omaggio a F. Benoit, Bordighera, 1971, V, 7-32, esp. 32.

48. T.W. Gallant, op.cit. (n. 39).

49. Naval Intelligence Division, Spain and Portugal (Geographical Handbook Series), II, London, 1941, 186. In general see R.J. Houk, 'The Portuguese fishing industry' in G. Borgstrom & A.J. Heighway (ed.), Atlantic Ocean Fisheries, London, 1961, 165-172.

50. Andalusia : F. Braudel, The Mediterranean and the Mediterranean World in the Age of Philip II (tr. S. Reynolds), London, 1972, 258; Malaysia : R. Firth, op.cit. (n. 5), 26-7; Polynesia : R. Firth, Primitive Polynesian Economy, London (2nd ed.), 1965, 112.

51. B. Didsbury, 'Cheshire saltworkers' in R. Samuel (ed.), Miners, Quarrymen and Saltworkers, London, 1977, 137-203, esp. 146, 150.

52. Geop. 20.46.3. 1 modius = 16 sextarii : see Lewis & Short, A Latin Dictionary, sv. modius & sextarius.

53. M. Ponsich & M. Tarradell, op.cit. (n. 26), 100-101. Salt was, however, shipped across the Mediterranean in the Roman period : Plut. Mor. 685d. In modern Malaysia Singapore is the major supplier of salt for all fish processing of the region and, thereby, has become the major entrepot of the fish trade : R. Firth, op.cit. (n. 5), 12.

54. See note 24.

55. 3.2.6 (144)

56. NH 31.39.73 - 31.45.105, where he first discusses salt, then garum and allex (first as foodstuffs and then as medicines) and finally salt as medicine.

57. Manil. Astron. 5. 656-92

58. 3.5.11 (175)

59. For Ibero-Punic amphorae at Alcacer : C.Tavares da Silva et al, 'Escavações arqueologicas no Castelo de Alcácer do Sal (campanha de 1979)' Setúbal Arq., 6-7, 1980-81, 149-218, esp. 186-7; for S. Olaia see A. dos Santos Rocha, 'Estações préromanas da idade do ferro nas vizinhanças da Figueira da Foz em Portugal', Portugalia, 2, 1905-1908, 301-359; 493-516.

60. E.N. Shanahan, op. cit. (n. 36), 363.

61. V. Rau, A exploração e o comercio do sal de Setúbal (= Estudo de historia económica, 1), Lisbon, 1951, esp. 24, 33-40. In more recent times Portugal produced such a surplus of salt that it could supply Newfoundland for fish-processing there : C.L. Cutting, Fish Saving : a history of fish processing from ancient to modern times, London, 1955, 129.

62. See V. Rau, op.cit. (n. 61), 22-3.

63. In general see J. Vila Valenti, 'Notas sobre la antigua producción y comercio de la sal en el Mediterraneo occidental', I Congreso arqueologico del Marruecos español, Tetuan, 1954, 225-234; cf. J. Alexander, 'The salt industries of Africa : their significance for European prehistory' in K. W. de Brisay & K.A. Evans (ed.), Salt : the study of an ancient industry (report on the Salt Weekend held at the University of Essex, 20-22 Sept. 1974), Colchester, 1975, 81-2.

64. For example, CIL III 1209 (Apulum, Dacia): conductor pascui, salinarum et commerciorum; CIL III 1363 (Veczel, Dacia) : conductor pascui et salinarum; ILLRP 734, 738, 743 (Minturnae, Italy) : socii salinatores. See in general M. Besnier, 'Sal' in C.V. Daremberg & E. Saglio (ed.), Dictionnaire des antiquités grecques et romaines (5 vols. in 9), Paris, 1873-1919, 4.2, 1009-1012.

65. CIL XI 390, 391, on which see E. Will, 'Le sel des Morins et des Ménapiens' in M. Resnard (ed.), Hommages à A. Grenier, III (= Collection Latomus, LVIII), Brussels, 1962, 1649-1657.

66. For Byzantium : Strabo 7.4.6 (311); J. Dumont, 'La peche du thon à Byzance à l'époque hellénistique', REA, 78-79, 1976-1977, 96-119, esp. 115.

67. Dig. 50.15.4.8; cf. P.A. Brunt, 'The revenues of Rome', JRS, 71, 1981, 161-72, esp. 167.

68. For details see J.A. Romero Magalhães, op.cit. (n. 38), 151-153.

69. F. Braudel, loc. cit. (n. 50).

70. J. Bourge, art.cit. (n. 41).

71. Dig. 43.14.1.7; Dig. 50.15.4.7

72. F.F. Abbott & A.C. Johnson, Municipal Administration in the Roman Empire, Princeton, 1926, 384-386.

73. Strabo 14.1.26 (642); the "telonion tes ichthuikes" was certainly in operation under Nero (OGIS 8858) and Antoninus Pius (OGIS 496).

74. J.A. Crook, Law and Life at Rome, London, 1967, 140.

75. IG 4 (2). 77, on which see P.E. Legrand, 'Inscriptions de Trézène', BCH, 24, 1900, 179-215, esp. 190-199; B. Haussoullier, 'Note sur une inscription de Trézène', Rev. de Phil., n.s., 25, 1901, 336-338. I am grateful to Dr. Gallant for drawing my attention to this inscription.

76. On the continued use of publicani under the Empire and communal tribute liability see : P.A. Brunt, art. cit. (n. 67), 168-9. M. Corbier, 'Fiscalité et dépenses locales' in P. Leveau (ed.), L'origine des richesses dépensées en la ville antique (Actes du Colloque à Aix-en-Provence, 11-12 mai 1984), Aix-en-Provence, 1985, 219-232, esp. 225-7.

77. For example at Tróia (C 35) : O. da Veiga Ferreira, 'Algumas notas acerca da pesca na antigüedade', AP, iii, 2, 1968, 113-133, esp. 130; for other *cetariae* of Sado estuary see A.I. Marques da Costa, 'Estudos sobre algumas estações da época luso-romana nos arredores de Setúbal', AP, 27, 1925-1926, 165-181, esp. 175-176. In general see J. Pedro da Costa, 'Estudo de fauna malocologica no espolio da gruta natural de Ibne Amar' *Actas do II Congresso Nacional de Arqueologia*, Coimbra, 1971, II, 599-616.

78. For detailed references RE XXIII.2, 2000-2020, sv. 'purpura'; M. Besnier, 'Purpura' in Daremberg-Saglio, 4.1, 769-779.

79. Plin. NH 9.62.133-134.

80. NH 9.62.133.

81. See above, p. 105-7.

82. See M. Besnier, art.cit. (n. 78). On Phoenicians in the West in general see A. Garcia y Bellido, op. cit. (n. 26); for more recent work and detail: J.M. Blazquez, *Tartessos y los origines de la colonización fenicia en Occidente* (2nd ed.), Salamanca, 1975, esp. 243-419; C.R. Whittaker, 'The western Phoenicians : colonization and assimilation', PCPhS, 20, 1974, 58-79; H. Schubart, 'Phönizische Niederlassungen an der Iberische Sudküste' in *Phönizier im Westen* (= Madrider Beitrage, 8), Mainz, 1982, 207-234.

83. A Phoenician colony : Str. 3.5.1 (168); on Punic amphorae produced there see J. Ramon, *La producción anfórica punico-ebusitana*, Madrid, 1981; seat of procurator : Not.Dign. *Occid.* XI. 71.

84. Fish-processing : M. Ponsich & M. Tarradell, op.cit. (n. 26), 82-89; purple : Str. 3.2.7 (145).

85. Fish-processing : M. Ponsich & M. Tarradell, op.cit. (n. 26), 7-77; purple shell middens : J. Desjacques &

P. Koeberlé, 'Mogador et les Îles Purpuraires', Hesperis, 42, 1955, 193-202.

86. On Mogador see J. Desjacques & P. Koeberlé, art.cit. (n. 85); A. Jodin, Mogador. Comptoir phénicien du Maroc atlantique (= Etudes et travaux d'archéologie marocaine, 2), Tangiers, 1966 and Les établissements du roi Juba II aux Îles Purpuraires (Mogador), Tangiers, 1967.

87. J.D. Evans, 'The exploitation of molluscs' in P.J. Ucko & G.W. Dimbleby (ed.), The Domestication and Exploitation of Plants and Animals, London, 1969, 479-484, esp. 479.

88. D.S. Reese, 'The industrial exploitation of murex shells : purple dye and lime production at Sidi Khrebish, Benghazi (Berenice)', Libyan Studies, 11th Annual Report of the Society for Libyan Studies, 1979-80, 79-93.

89. For price see S. Mrozek, 'Le prix de la pourpre dans l'histoire romaine' in Les dévaluations a Rome. Époque républicaine et impériale II (Gdansk 19-21 oct. 1978) (= Coll. de l'école française de Rome, 37), Rome, 1980, 235-43.

90. Strabo 3.2.6 (144); Plin. NH 8.7.

91. J.-G. Gorges, Les villas hispano-romaines : inventaire et problématique archéologiques, Paris, 1979, sv PS 31 & p. 479.

92. See Plin. NH 31.44.95.

93. J.D. Evans, art.cit. (n. 87), esp. 480; J. Roche & O. da Veiga Ferreira, 'Les fouilles récentes dans les amas coquilliers mésolithiques de Muge (1952-1965)', AP, iii, 1, 1967, 19-41; cf. O. da Veiga Ferreira, 'Acerca da presença de "Purpura haemastoma" e "Purpura lapillus" Linné nas estações pré-historicas portuguesas', Rev.Guim., 68, 1958, 377-382 for the view that they were exploited for their purple dye. For modern use see photographs of market stalls in Marseilles and Venice at D.S. Reese, art.cit. (n.

88), Pls. 2 & 3.

94. Rotura : O. da Veiga Ferreira, art.cit. (n. 3), 123-124; Pedrão : J. Soares & C. Tavares da Silva, 'Ocupação do periodo proto-romano do povoado do Pedrão (Setúbal),' Actas das II Jornadas Arq., 1972, I, 245-306, esp. 266; Santa Olaia : A. dos Santos Rocha, art. cit. (n. 59), esp. 354.

95. M.R. Jarman, G.N. Bailey & H.N. Jarman (ed.), Early European Agriculture : its foundation and development, Cambridge, 1982, 74. I am also grateful to Dr. T.W. Gallant for information on this point. The twofold use of Murex and Purpura shell-fish is perfectly captured by Mart. 13.87.

96. See further M. Ponsich & M. Tarradell, op.cit. (n. 26), 9-37.

97. ibid., 55-68. I prefer the term "cocciopesto" to the somewhat confusing "opus signinum" favoured in most accounts of these tanks.

98. ibid., 40-55.

99. For a discussion of the significance of the physical layout of villas for the nature of the community attached see J.T. Smith, 'Villas as a key to social structure' in M. Todd (ed.), Studies in the Romano-British Villa, Leicester, 1978, 149-185 and 'Villa plans and social structure in Britain and Gaul', Caesarodunum, 17, 1982 (= Actes du colloque : La villa romana dans les provinces de nord-ouest), 321-331.

100. 20.46.1-2

101. 20.46.3

102. 20.46.5

103. See Chapter Four, pp. 77-81.

104. D.P.S. Peacock, _Pottery in the Roman World : an ethnoarchaeological approach_, London, 1982, 38-43; see also M.G. Fulford, 'The location of Romano-British kilns : institutional trade and the market' in J. Dore & K. Greene (ed.), _Roman Pottery Studies in Britain and Beyond_ (BAR Suppl.Ser., 30), Oxford, 1977, 301-316, esp. 307.

105. G. Riccioni, _La terra sigillata della Gallia meridionale_ (= Problemi di archeologia gallo-romana, I), Bologna, 1977, 107.

106. See D.P.S. Peacock, op.cit. (n. 88), 122; F. Favory, 'Le monde des potiers gallo-romains', _Les dossiers de l'archéologie_, Sept. - Oct. 1974, 90-102; H. Vertet, 'Pauvres potiers, pauvre misère', ibid., 85-89.

107. On _societates_ in mining see Chapter Four, p. 71-2; on _societates_ in salt-working, see above p. 116 & note 64.

108. G. Riccioni, op.cit. (n. 105), 108.

109. D.P.S. Peacock, op.cit. (n. 104), 123.

110. G. Pucci, 'La ceramica italica (_terra sigillata_)' in A. Giardina & A. Schiavone (ed.), op. cit. (n. 19), 99-121, esp. 102-6.

111. Carlos Ribeira, reproduced by A.I. Marques da Costa, art.cit. (n. 77), 166-172, esp. 169.

112. D.F. de Almeida, J. & A. Cavaleiro Paixão, _Notas sobre a estação arqueológica de Tróia de Setúbal_, Setúbal, 1978, 2.

113. R. Firth, op.cit. (n. 5), 10.

114. Chapter Four, pp. 71-81.

115. As occurs in modern fish processing in the Third World : see R. Firth, op.cit. (n. 5), 8-9.

116. For a parallel connexion of villas and fish-processing in north-west Gaul see J.F. Drinkwater, Roman Gaul, the three provinces 58 B.C. - A.D. 260, Beckenham (Kent), 1983, 162.

117. A. Grenier, Manuel d' Archéologie gallo-romaine. II.2 : Navigation, occupation du sol, Paris, 1934, 628.

118. Dig. 33.7.25.

119. P.S. Middleton, Trade and Market in Roman Imperial Gaul (unpubl. Ph.D. thesis), Cambridge, 1982, ch. 4.

120. H. Cockle, 'Pottery manufacture in Roman Egypt : a new papyrus', JRS, 71, 1981, 87-97, esp. 92.

121. M. Ponsich & M. Tarradell, op.cit. (n. 26), 98.

122. A.I. Marques da Costa, 'Estudos sobre algumas estações da época luso-romana nos arredores de Setúbal', AP, 29, 1930-31, 2-31, esp. 19-31.

123. M. Maia, 'Contribuição para o estudo das ánforas romanas de Tróia - ánforas do tipo "Africano Grande"', Setúbal Arq., 1, 1975, 155-158.

124. For names in -ianus suggesting freedmen : A.M. Duff, Freedmen in the Early Roman Empire, Oxford, 1928, 53; for "Greek and Oriental names" P. Garnsey, 'Descendants of freedmen in local politics : some criteria' in B. Levick (ed.), The Ancient Historian and his Materials : studies in honour of C.E. Stevens on his seventieth birthday, Farnborough (Hants.), 1975, 167-180, quoted from p. 173.

125. See M.I. Finley, The Ancient Economy, London, 1973,

64; S.M. Treggiari, *Roman Freedmen during the late Republic*, Oxford, 1969, 109-115.

126. R. Firth, op.cit. (n. 5), 10.

127. J.H. d'Arms, *Commerce and Social Standing in Ancient Rome*, Cambridge (Mass.), 1981, 121-148; S.M. Treggiari, op.cit. (n. 125), 91-106.

128. P.D.A. Garnsey, 'Independent freedmen and the economy of Roman Italy under the Principate', *Klio*, 63, 1981, 359-371.

129. For some suggestive parallels from the organization of production of later Italian *terra sigillata* see G. Pucci, art. cit. (n. 110), 114-5.

130. M. L. Abreu Nunes, 'Tesouro de moedas romanas encontradas em Tróia', *AP*, iii, 7-9, 1974-77, 359-364; for a parallel from a Baetican fish-processing site : F. Molino Fajardo & J.L. Lopez Castro, 'Conjunto de monedas de la factoria de salazones de "El Majuelo"', *Almuñecar. Arqueologia e Historia*, 2, 1984, 257-74.

131. For fishing as the equivalent of a cash crop see T.W. Gallant, op. cit. (n. 39), esp. 40-42.

132. B. Liou & R. Marichal, 'Les inscriptions peintes sur amphores de l'anse Saint-Gervais à Fos-sur-Mer', *Archaeonautica*, 2, 1978, 109-181, esp. 131-135 on a *titulus pictus* : SCOMB(ri) / FLOS / AA / (.) IUNI CILONIS / (E)X OF. AUGG on Dressel 7-11 amphora from Baetica.

CHAPTER SIX

COMMERCE

i. Ports, Traders and Shippers

Little detailed work has to date been done on Lusitanian ports. As already suggested, the main port of the province in the Principate was Olisipo, as a result of which Salacia lost most of its former importance. Archaeology has not as yet revealed any port installations at Olisipo, but the recent discovery of a Roman garum factory beneath the Casa dos Bicos with an associated small quay may eventually lead to further discoveries in this zone. (1) The Algarve was served by Portus Hannibalis and Ossonoba; as regards the former, in addition to its name, the discovery of a Roman anchor in the estuary of the Arade (see Figure 6.1) is the only indication of its function as a port; for the latter, two pieces of evidence may be indicative, but their interpretation remains hopelessly controversial. The first consists of an almost illegible inscription on a copper ingot that some have reconstructed as recording a "procurator Ossonobensis". The role of such a procurator may have been to control the shipment from Ossonoba of the mineral resources of the Algarve and lower Alentejo, essentially copper and silver, as we have seen. (2) Furthermore, he may have had a more general role in supervising the import and export of produce from the port. Secondly, a mosaic has recently been discovered in Faro, with a huge central emblema depicting a personified Oceanus, surrounded by four winds one at each corner. Beneath is an inscription, which has been interpreted as revealing that the mosaic was laid by the magistri of a corporation; given the nautical theme of the mosaic, it may have been a corporation of shippers operating out of Ossonoba. (3) Although this interpretation of the mosaic

FIGURE 6.1 ROMAN ANCHORS FOUND OFF LUSITANIA

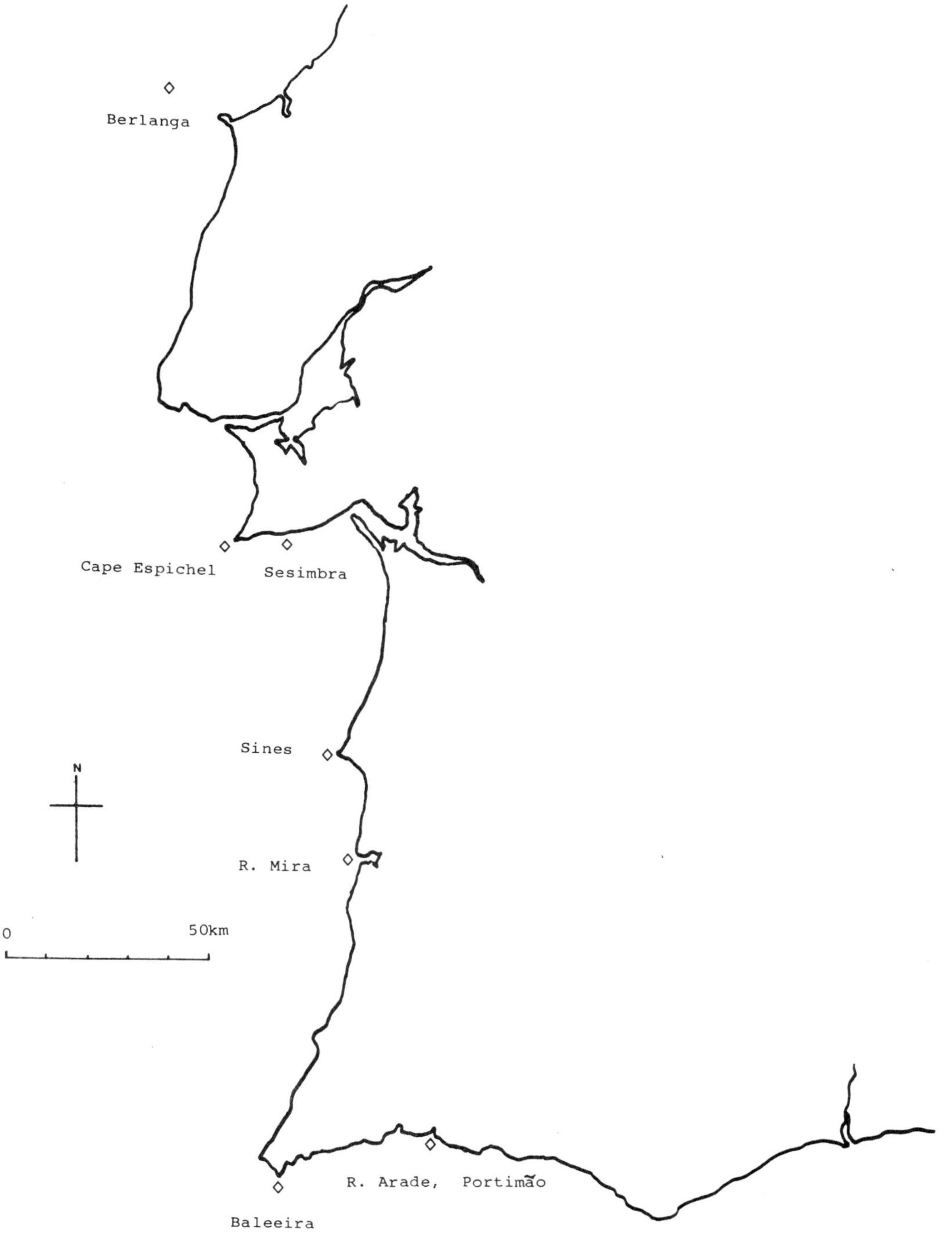

is seductive, the significance of Ossonoba as a port remains largely still to be proven.

These were not the only ports of the province. Myrtilis (Mértola) was an important river port on the Anas (Guadiana), while Aeminium (Coimbra) on the Mondego and Scallabis (Santarém) on the Tagus seem likely to have been others. Some have held that Emerita, some three hundred kilometres up the Guadiana, also had a river port and have used the archaeological remains still to be seen on an island in the middle of the river as confirmation of this. However, they ignore the problem of the rapids and the sheer drop in the river at Pulo do Lobo; it is thus impossible to sail from Emerita to the mouth of the Guadiana. But this does not preclude river traffic at least between Emerita and Badajoz. (4) Setúbal itself must have had port facilities, while smaller ports perhaps existed at Figueira da Foz at the mouth of the Mondego, at Sines, serving Mirobriga (modern Santiago do Cacem) and at Lagos (?Lacobriga), Balsa (Tavira) and Baesuris (São Bartolomeu de Castro Marim) in the Algarve. Not surprisingly at Troia remains of a Roman quayside can still be seen near the so-called "Rua da Princesa". (5)

The evidence for Lusitanian traders in general is scant and for Lusitanian fish-sauce traders in particular almost non-existent. (6) The surviving amphorae which carried such products rarely bear a revealing stamp or *titulus pictus*. That there were specialized traders in fish products is not in doubt. Inscriptions from various parts of the Empire attest such specialists as *negotiatores salsarii* (7), *negotiator muriarius* (8) or *negotiatores allecarii* (9). Occasionally a trader diversified by selling both fish products and wine : thus the *negotiator salsamentarius et vinarius* at Rome. (10)

The only evidence that can throw any light on the question consists of two inscriptions, which both record the careers of *codicarii*. Contrary to the belief of some, the vessels run by *codicarii* were not necessarily just river barges, but boats which are said to have "carried an effective spread of sail and were able to be used for coastal as well as harbour and river work". (11) Thus, in Lusitania such boats could have been used not only to transfer products from the centres of production to the ports, where larger vessels could in turn ship them further afield, but also to transfer products to Gades (Cadiz), indisputably the major port of the western Iberian peninsula. The provenance of both inscriptions is significant. One comes from Olisipo, the major Atlantic port of the province, the other from Mirobriga Celtica, near the coast some fifty kilometres south of the Sado estuary. Thus, coastal transfer of surplus produce is attested in just that part of the province where fish-processing was most prevalent.

Moreover, these inscriptions also reveal something of the social status of such shippers. Both were men of high social standing within their local communities : L. Iulius Maelo was flamen divi Augusti at Olisipo under Tiberius, (12) while M. Iulius Marcellus was aedile and II vir at Mirobriga. (13) Their holding of such prominent positions would imply that they had a considerable basis of wealth, probably derived from landed estates held within the territories of their respective towns. They had presumably invested some of this wealth in shipping. Whether they were also involved in the production of fish sauce is impossible to ascertain.

Epigraphic evidence from Rome suggests that there codicarii were closely associated with navicularii and, therefore, likely to have been men of similar social status. (14) It is now thought that men of the highest social position invested some of their wealth in shipping, especially since some of the merchantmen required a large investment of capital because of the size of operation involved. (15) Various navicularii involved in the Baetican olive oil trade were prominent men in their local communities : for example, P. Olitius Apollinaris, a freedman who rose to the prestigious position of VI vir Augustalis at Narbo in Gaul; more striking still is Sex. Fadius Secundus Musa, who held all the public offices of Narbo, and Q. Fulvius Carisianus, who became pontifex (the highest local honour) in Arva in Baetica. (16) The evidence for shippers from Lusitania, scant though it is, would seem to confirm the hypothesis that shippers were often men of high status and wealth in their local communities.

The attestation of codicarii, rather than navicularii, could be mere chance, but may reflect the nature of Lusitanian shipping. Although ports did exist in Lusitania, there is nothing yet to suggest that they ever developed into major ports serving the Mediterranean basin on the same scale as Gades or Carthago Nova (Cartagena). It seems more likely that they served as entrepots, whither surplus produce was transported from the interior and coastal regions of the province, thence to be transferred in small vessels to the nearest major seaport, Gades in Baetica. Strabo, writing under Augustus and Tiberius, but sometimes reliant on sources of the first century B.C., gives a hint of Gades' interest in the Atlantic shipping route when he says that "the people of Gades fit out the most and the largest ships both for our sea (i.e. the Mediterranean) and for the sea outside (i.e. the Atlantic)". (17) Thus, a system of cabotage operated between Lusitania and the major ports serving the central Mediterranean.

In certain areas of the Roman Empire (notably southern France) underwater archaeology has added considerably to our knowledge, and understanding of Roman shipping and trade routes. Unfortunately, very little underwater work has to date been carried out in Portuguese waters : not one Roman ship has yet been investigated. But various anchors have come to light and their findspots (see Figure 6.1) seem to indicate Roman ships plying the Atlantic along routes directly linked to the fish-processing industry. Although none have yet been found in the Sado estuary, a relatively high concentration occurs just to the south. In fact so many anchors have been found in the Mar de Ancão south of Cape Espichel, some fifteen plus, that it has been plausibly suggested that this was an anchorage site for ships not wishing to enter the Sado estuary. (18) With the exception of the anchor from the mouth of the Rio Mira, all have been found in proximity to cetariae.

However, another trade route is beginning to emerge from amphora finds and submarine archaeology in the north-west of the Iberian peninsula. It now seems clear that some produce was shipped north from Lusitania and Baetica to the Atlantic coasts of Spain, France and possibly Britain, but the extent and scale of this cannot yet be ascertained. (19) The Roman lighthouse still standing at La Coruña, built by a Lusitanian architect from Aeminium (Coimbra), provides further testimony. (20)

ii. Lusitanian fish-sauce amphorae

Another more profitable angle of approach is to examine the containers in which Lusitanian fish products were shipped, namely the amphorae. It remains a possibility that other types of ceramic containers or even barrels were also used, but it is impossible to assess their role relative to that of amphorae because the former are more difficult to categorize than amphorae and the latter have simply not survived. A Roman ship could carry a cargo of greater capacity, if barrels were used rather than amphorae. (21) Furthermore, salted tunny was shipped in barrels from the Algarve in the Medieval period. (22) But given that a suitable and ample supply of clay existed near most production centres in Lusitania, it is likely that amphorae played a more significant role in the shipping of the province's surplus.

Any study of Lusitanian amphorae must begin with the kilns in which they were produced. This is the only secure basis for attributing a Lusitanian origin to a particular form of amphora. Once the specifically Lusitanian products can thus be identified, it may then be possible to trace their diffusion within the Iberian peninsula and further afield. This in turn may then allow some impression to be gained about the horizons of the exchange of Lusitanian fish-products and the historical development of this exchange.

An investigation of this type is beset with problems of evidence. It is only recently that amphorae have received any scholarly attention. Typologies are becoming ever more refined, but are still not sophisticated enough to take in slight local variations. (23) It remains a contentious issue whether all amphorae of the same general type were produced in just one particular region. Peacock is surely right to argue that not all the supposedly "Baetican" types of fish-sauce amphora were necessarily produced in the kilns of Baetica. Imitations may well have been produced in the neighbouring kilns of southern Lusitania. (24) Conversely it follows that amphorae identified here as

"Lusitanian" may also have been produced elsewhere. Thus it is now becoming clear that arguments based only on the shape and typology of amphorae are not sufficient for a secure identification of their place of origin, but must be complemented by a petrological analysis of the fabric of the amphora itself. (25) Such analysis is only beginning to be given to Lusitanian amphorae. What follows should, therefore, be treated as a provisional, exploratory picture. It is, however, important to provide some sort of reconstruction at this stage, which can then be confirmed or modified in the light of subsequent discoveries.

Another complication is introduced, if it is accepted that some amphorae were produced in one area for filling in another. Pottery leases from Oxyrhynchus in Roman Egypt make it clear that the vessels manufactured are to be filled with wine produced on various other estates. (26) But, on the whole, if a good supply of clay existed in the fish-processing areas, why would a producer have preferred to import empty amphorae rather than use locally manufactured ones ? Especially if the producers were also landowners (as I have argued they often were), they would doubtless have had the amphorae manufactured by their rural peasants on their landed estates.

What is immediately striking is that a large majority of those Lusitanian amphora kilns yet discovered are located in those very parts of the province where fish-processing was practised, namely the Algarve and the Sado estuary (see Figure 6.2). On the whole the kilns in the Algarve were discovered at the end of the last century, whereas those in the Sado estuary have only recently been investigated. The evidence for the latter is, therefore, better documented and more sophisticated, with some petrological analysis of the amphorae undertaken. It is, therefore, on this area that the following discussion will largely concentrate.

Of the kilns in the Sado estuary, those at Quinta da Alegria and Herdade do Pinheiro seem to have produced amphorae for a longer period than the others. They have

FIGURE 6.2 AMPHORA KILNS OF LUSITANIA

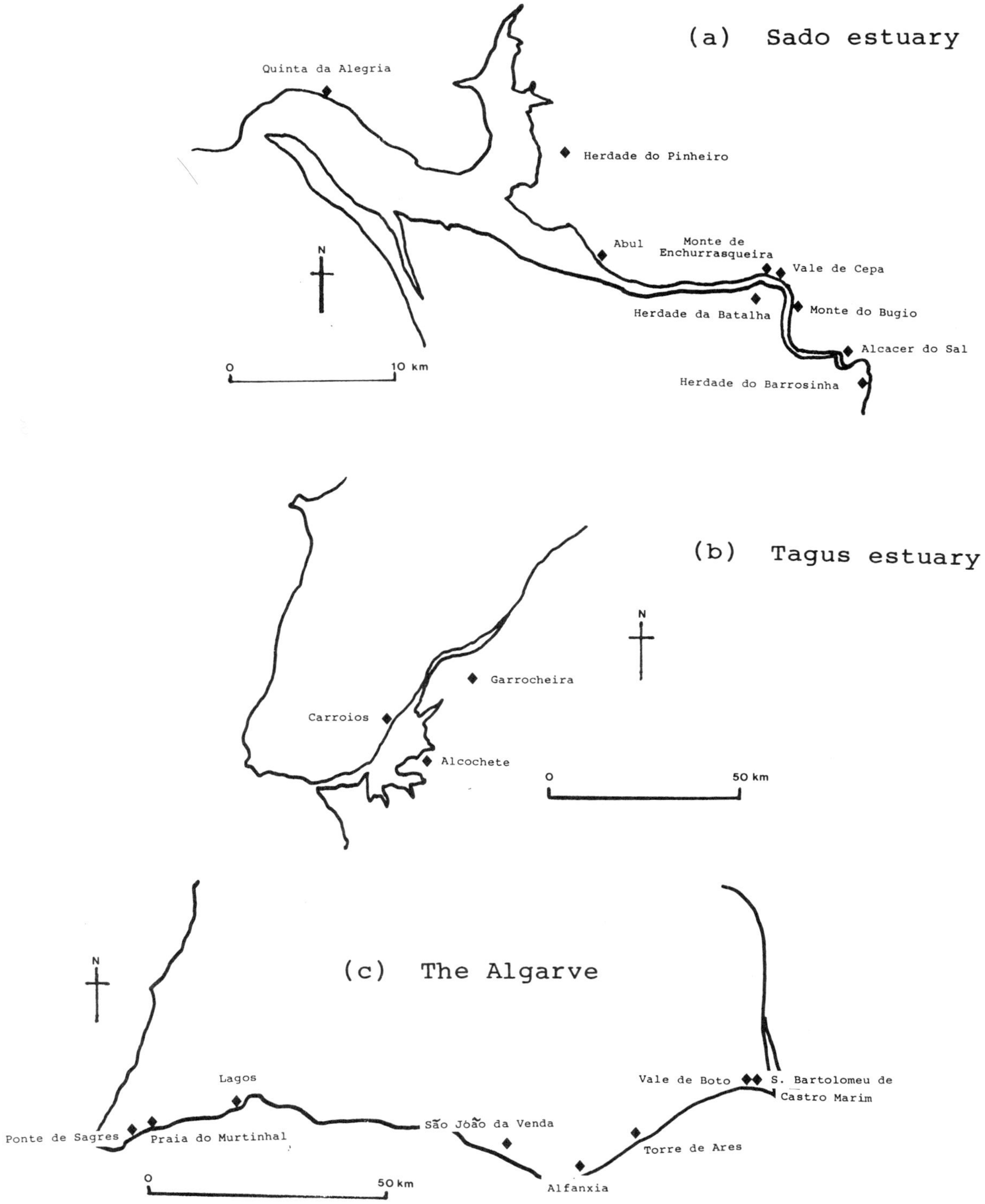

received more attention, however, and so more types can be identified. With further work at the other kilns, more types may be identified, which will demand a reconsideration of their historical development. It will be useful to list the kilns thus far discovered in Lusitania and give some rough guide to what each kiln produced.

It is quite clear that there were three main types of amphora being produced in these kilns : namely, those traditionally referred to as Dressel 14/Beltran IVb, Almagro 50 and Almagro 51c. Of these the first seems to have been the most common; to date only two kilns can be shown to have produced Almagro 51a-b; other local types were also relatively widespread. It is also clear that most amphora kilns also produced other types of coarseware : amphora lids, tiles, bricks and other construction materials, dolia and smaller pots and vessels.

Since the names for these amphora types were invented as part of much larger amphora typologies, and since they remain essentially cryptic, I propose to rename them as follows for ease, and clarity, of reference :

Dressel 14/Beltran IVb	=	LUSITANIAN GARUM I
Almagro 50	=	LUSITANIAN GARUM II
Almagro 51a-b	=	LUSITANIAN GARUM III
Almagro 51c	=	LUSITANIAN GARUM IV.

The first named was essentially produced in the late first and second centuries A.D., while the second and third types were current from the third to the later fourth/early fifth centuries. (27) The types seem not to have been exclusive to the Sado estuary, since very similar types to the Lusitanian Garum I were produced at the kiln of São Bartolomeu de Castro Marim. (28) There were other types of amphora produced in the Sado estuary, but only for local use, since they are not found outside their production area.

FIGURE 6.3 LUSITANIAN GARUM I AMPHORAE

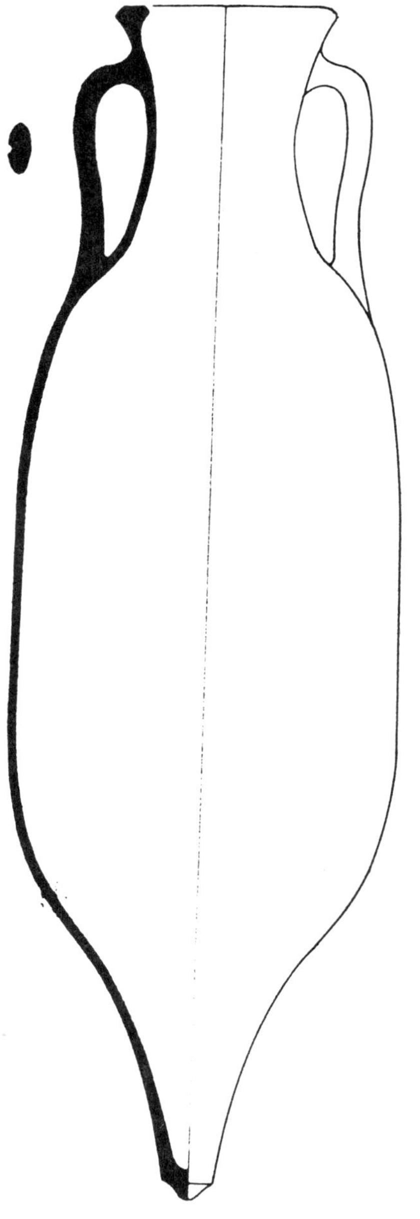

(after G. Cardoso, Conimbriga, 17, 1978, fig. VI, no. 12; scale 1:6)

A. <u>EARLY EMPIRE : FIRST TO EARLY THIRD CENTURY A.D.</u>

<u>LUSITANIAN GARUM I</u> (= DRESSEL 14/BELTRAN IVb)

(cf. Figure 6.3)

The contents for each particular type of Roman amphora are often disputed, but in this case there is little doubt that Lusitanian Garum I were used for the transportation of fish products. First, this is <u>a priori</u> likely from the location of the kilns and the economic pattern of the area. Secondly, it is confirmed by traces of fish-sauce that have been found in such amphorae. (29) Another amphora of this type from the Saint-Gervais wreck corroborates the hypothesis. Of the very same clay as those amphorae found in the kilns themselves, it bears the <u>titulus pictus</u> : (30)

LIQ(uamen)
EXC(ellens)
SABINI ET AVITI.

This form is stylistically derived from the more widely produced Beltran IV type, now shown to have been fabricated at the kilns of Calahonda (Granada) and Motril in Baetica during the first century A.D. (31) By the end of the first century, this variant form (Lusitanian Garum I) had started to be produced in the kilns of the Sado estuary. From associated fineware finds the kiln of Quinta da Alegria can be shown to have been producing this amphora form from c. A.D. 50. (32) Other finds of the type from Lusitania and elsewhere seem to confirm this. In the Praça do Bocage, Setúbal, six examples were discovered in a level with a <u>terminus ante quem</u> of c. A.D. 75; the Herculaneum example provides a similar <u>terminus ante quem</u> of A.D. 79; at Conimbriga fragments appear in Claudian levels. But the earliest example to date comes from the shipwreck Sud Lavezzi 3, which is dated to the first decades of the first century A.D. It is the only example of the type on board and so should perhaps be seen as part of the crew's provisions rather than as an item of trade.

This was perhaps not the first type of amphora to be produced in the kilns of the Sado estuary; for certain Dressel 7-11/Beltran I <u>garum</u> amphorae recently excavated in the castle at Alcácer do Sal are of the same Sado estuary clay as the Lusitanian Garum I examples found on the same site. The suggestion that Beltran I amphorae were produced from the end of the first century B.C. until they were supplanted by Lusitanian Garum I types in the mid- first century A.D. seems plausible. (33) But unless it can be shown from the stratigraphy that the Lusitanian Garum I types were found in later levels than the Beltran I types, what is to say that the two forms did not coexist for a while, until the former came to predominate ? After all both forms occur in a level of the last quarter of the first century A.D. in the Praça do Bocage, Setúbal. Perhaps then future excavations will confirm that there were two types, rather than one, of early Lusitanian amphora, both for the transport of fish products and both imitations of Baetican prototypes.

As for the later history of this type, it continued to be produced until sometime in the third century A.D. Most finds can be dated to the first, or more commonly, the second century A.D. However, some later finds do occur. In the Baths of the Swimmer, Ostia, it is found in Severan levels, while in the Praça do Bocage, Setúbal, it is found in destruction levels of the third century together with those other local types of amphora, Lusitanian Garum II and IV, which came to replace it. In the latter case there is a possibility that it is residual and production had ended some time before its deposit. But the fact that so many more kilns produced Lusitanian Garum I as opposed to Lusitanian Garum II, III and IV may be explained on the grounds that some kilns continued to produce only the former type, while others went over to the new types. On this argument there was some overlap or coexistence in the first half of the third century of types that have usually been exclusively labelled "early" or "later" Empire.

In his recent survey of Lusitanian amphorae Parker has put much emphasis on the appearance of the Lusitanian Garum I : he has suggested that this indicates a shift in the production of fish-sauces from Baetica to Lusitania at the end of the first century A.D. (34) There are some problems over his interpretation of the material. First, his dating of the appearance of the Lusitanian variant now seems rather late. As we have seen, the middle, rather than the end of, the first century A.D. seems more satisfactory. Secondly, such a sudden decline in Baetica's share of the market is difficult to accept without further demonstration. Baetican products could, after all, have continued to have been exported, but in a different form of amphora. However, the appearance of a distinctively Lusitanian form of amphora in the mid-first century A.D. fits very neatly into the general chronology established earlier for the Lusitanian fish-processing industry. Most fish-processing sites seem not to have come into operation until the end of the first century B.C. at the earliest. It would have taken some time for the production centres to take root and develop. In this adolescent phase the products of the industry were probably shipped in amphorae imported from Baetica or, more likely, imitated directly from Baetican models, if indeed there was a surplus to be shipped. Moreover, it would have taken a generation or so for the province of Lusitania to become integrated within the economic system of the Empire. Once this had occurred, the products of the province may then have been able to compete with, to complement, or in time perhaps to eclipse the products of Baetica.

B. LATER EMPIRE : THIRD TO FIFTH CENTURY A.D.

At the end of the third century A.D. certain kilns in the Sado estuary and the Algarve started to produce three new forms of amphora, which after a period of coexistence came to supercede the Lusitanian Garum I type, namely the types Lusitanian Garum II, III and IV. That all these amphorae were also destined for fish products is not in serious doubt, once again given the location of the kilns and the discovery of fish remains inside such amphorae. (35)

There were also slightly variant forms of these types of amphora. One of the merits of Keay's recent typology of later Empire amphorae is that he differentiates between two variants of the Lusitanian Garum II type (his Types XVI and XXII). (36) Variants of the Lusitanian Garum IV type also occur : most notably Parker's "Lusitanian Piriform". (37) The recent discovery of this type at the kilns at Quinta da Alegria and Herdade do Pinheiro, its petrological analysis and its distribution limited merely to the province of Lusitania all point irrefutably to a Lusitanian origin. Similarly Cardoso has identified a type which he designates "anfora lusitana", but which is surely a variant form of a Lusitanian Garum IV type amphora. (38) Both Lusitanian Garum IV and "anfora lusitana" types are found together at the kiln at Vale do Boto. (39)

The chronology for, and production of, each of these later types is best discussed separately.

i. LUSITANIAN GARUM II (= ALMAGRO 50)

(cf. Figure 6.4)

Keay has done a useful service in collecting the data for the chronology of this type. (40) Luni provides the earliest examples - from the later second century A.D., but fragments are more common from levels of the early third century. The Ognina wreck, and thus also the Lusitanian Garum II amphorae on board, has been dated tentatively to A.D. 210-215, while the Monte Testaccio fragments must date from earlier than the mid third century, when the dump went out of use. Thus Lusitanian Garum II amphorae seem to have been produced from the end of the second century A.D. They then continue right through into the fifth century, as the carefully dated finds from the Baths of the Swimmer, Ostia, and the wreck Anse Gerbal demonstrate. Fragments have been found even in sixth century levels at Pollentia, Mallorca, but, as Keay notes, these were probably residual.

That these amphorae were produced in the Sado estuary is not in doubt, given that examples have been found in the kiln sites themselves and analysis of their clay confirms their Sado origin. But whether they were also produced in the Algarve remains something of a problem. To date, fragments of this type have only come to light from the kiln at São Bartolomeu de Castro Marim, and furthermore, only in small quantities. This raises the question of whether they were actually produced in these kilns or imported. Unfortunately no clay analysis has yet been done to help settle this question. However, we are fortunate that unlike most other Lusitanian amphorae Lusitanian Garum II were occasionally stamped. The few

FIGURE 6.4 <u>LUSITANIAN GARUM II AMPHORAE</u>

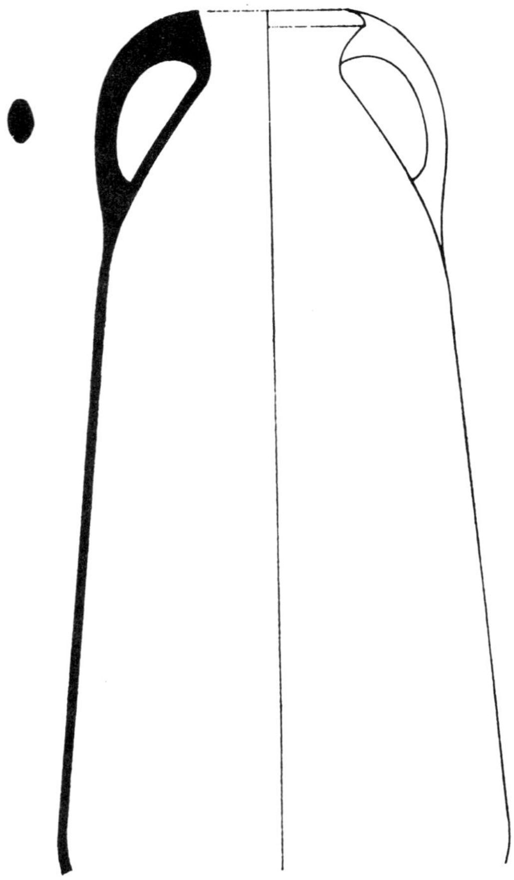

(after G. Cardoso, <u>Conimbriga</u>, 17, 1978, fig. X, no. 22; scale 1:6)

stamps that survive add weight to the hypothesis that these amphorae were produced in the Algarve, and more specifically at the kilns at Balsa (Tavira). For of the four commonest stamps examples of three have been found in the territory of Balsa (see Figure 6.5). In addition to its appearance on amphorae, the stamp AEMHEL also appears on construction materials (tegulae and imbrices) which have been found at Quinta de Torre de Ares and Quinta das Antas. This provides a vivid illustration of the variety of ceramic artefacts that were produced at a single kiln. (41) The same stamp has recently been discovered on an amphora from the island of Pessegueiro. The fabric of the amphora suggested to the excavators an origin in the kilns of the Sado estuary. If this is correct, it raises the strong possibility of a firm of potters which operated branch workshops. (42)

FIGURE 6.5

POTTERS' MARKS ON LUSITANIAN GARUM II AMPHORAE

(A) LEV. GEN. = L. Eu.... Gen(ialis) : S. Keay, 151

1.	BALSA, Quinta de Torre de Ares	*AP*, 12, 1907, 226; *AP*, iii, 2, 1968, 100.
2.	Estói, Milreu (nr. Faro) (Roman villa)	kind information from Prof. Hauschild
3.	ITALICA (nr. Seville)	*MEFRA*, 90, 1978, 382
4.	OSTIA or Rome	*CIL* XV 3452,2
5.	ALBA FUCENS	*NSA*, viii, 6, 1952, 236
6.	?? (now in Museum at Rabat, Morocco)	*MEFRA*, 90, 1978, 382 & Pl. XIII,1

7.	SAMARIA, Palestine	J.M. & J.W.Crowfoot & K.H. Kenyon, _The Objects from Samaria_, 1957, 388, no. 4

(B) AEMHEL or AEMIHEL = ? Aem(ilius) Hel(iodorus) : Keay

1.	BALSA - Quinta de Torre de Ares	_AP_, iii, 1974-77, 250 & figs. 12-15
2.	BALSA - Quinta das Antas	_CIL_ II 6254, 2
3.	Island of Pessegueiro, Sines	_Arquivo de Beja_, ii, 1, 1984, 19
4.	? NARBO (museum at (Narbonne)	_Études sur Pézenas et l'Hérault_, 11, 1980, 7 & note 21
5.	ROME	_CIL_ XV 3394 a, b

(C) OLYNT = Olynthus

1.	BALSA - Quinta de Torre de Ares	_CIL_ II 6254, 33
2.	OSTIA - Baths of the Swimmer	_Ostia_ IV, 121 & Tav. XXXIV, fig. 254
3.	LUNI	_Scavi di Luni_, 1974, 455

(D) Q.E.TAVR. (or TAVR. or Q.E.T.)

1.	ITALICA	_Études sur Pézenas et l'Hérault_, 11, 1980, 7 & note 21

2.	?? (museum of Narbonne)	CIL XII, 5683, 88
3.	OSTIA (Baths of the Swimmer)	Ostia IV, 121 & Tav. XXXIV, fig. 253
4.	ROME (Monte Testaccio)	CIL XV 3448
5.	LUNI	Scavi di Luni, 1974, 455 & Pl. 111, 10

(E) OTHERS

I. ANCE.

1.	OSTIA (Baths of the Swimmer)	Ostia III, 613
2.	Roselle, Etruria	Ostia III, 606

II. CVRVCVNTIN.

1.	OSTIA (Baths of the Swimmer)	Ostia III, 606

III. M.

1.	LACOBRIGA (Lagos)	A.J. Parker, Méthodes classiques et formelles, 1977, fig. 23
2.	EMPORIAE	S. Keay, Later Roman Amphorae, 1984, 150

IV. MA

1.	LACOBRIGA	A.J. Parker, ibid., fig. 24

V. OPIL.

1.	BARCINO	S. Keay, ibid., 150.

VI. SVI (on 1 handle) VIC (on the other)

1. EMPORIAE — S. Keay, ibid., 150 & fig. 207.4

As yet no example of the mark Q.E.TAVR. has been found in the Algarve or the Sado estuary, but the very fact that it was stamped on a Lusitanian Garum II amphora suggests that it was a Lusitanian mark.

ii. <u>LUSITANIAN GARUM III</u> (= ALMAGRO 51 a-b)

(cf. Figure 6.6)

Keay has again collected the data on the chronology of this type and concluded that it was produced from the late third century until the later fifth century. (43) The recent excavation at the Schola Praeconum in Rome has revealed an example of this type in a level of 430-440. The Sado estuary was perhaps not the only area that produced this type of amphora, since one has been found in a fish-sauce establishment near Cartagena (44) and finds have been more common along the coasts of Baetica and Hispania Tarraconensis rather than in Lusitania. But Keay's inventory list excludes recent finds from the kilns of Quinta de Alegria and Herdade do Pinheiro, from Alcacer do Sal and from the island of Pessegueiro off Sines. Until more work is done on the clay fabrics of these amphorae, and on this type in general, it is difficult to say if the finds from Baetica and Tarraconensis represent items exported from Lusitania or local products.

FIGURE 6.6 <u>LUSITANIAN GARUM III AMPHORAE</u>

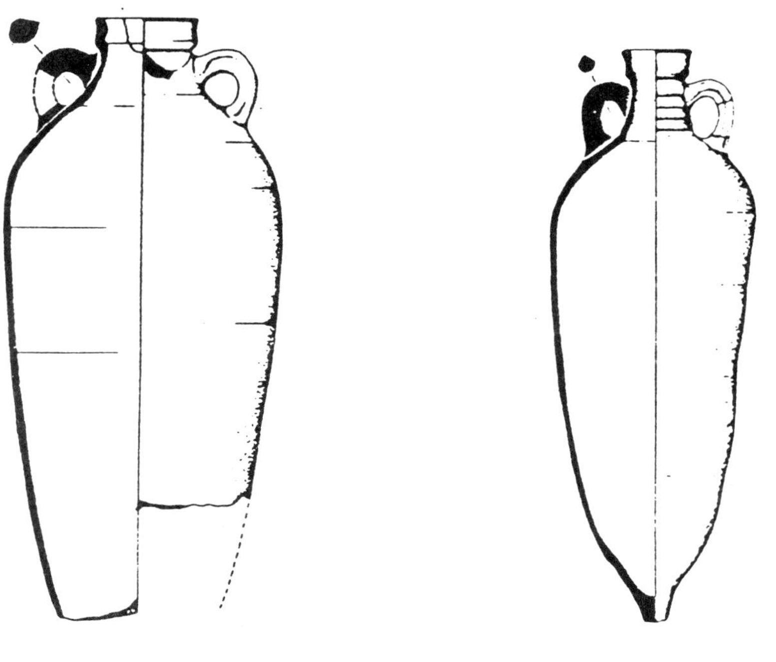

(after S.J. Keay, <u>Later Roman Amphorae</u>, 1984, fig. 21, nos. 8 & 10; scale 1:10)

iii. <u>LUSITANIAN GARUM IV</u> (= ALMAGRO 51c)

(cf. Figure 6.7)

Of all types produced at Quinta da Alegria, Lusitanian Garum IV amphorae have been found in the greatest abundance in the locality of the kiln, some 56.3 per cent. (45) There is also secure evidence that they were produced in the Algarve. The earliest datable example is that from the western cemetery of Caesarea, Tipasa (Mauretania), dated to the second century A.D.; but most examples come from contexts datable to the third or fourth centuries. Thus this type occurs in levels of the second quarter of the third century at the Baths of the Swimmer, Ostia and on the Femina Morta wreck, dated to the fourth century. It remained in production into the fifth century, since it occurs in the Anse Gerbal wreck and the Palaeochristian cemetery at Tarragona.

The evidence from the Baths of the Swimmer, Ostia is so extensive that it allows an interesting picture to emerge of the history of the various fish-sauce amphorae from southern Spain in general and Lusitania in particular. For Lusitanian Garum II and IV types of amphora can now be seen to have replaced other kinds of fish-sauce amphora (both Baetican Beltran II and IV and Lusitanian Garum I) in the later Severan period. As has been noted, they then continue to be found in levels of the end of the fourth century, and even the early fifth century A.D. However, in comparison with other types of amphora that carried various other foodstuffs, they are not present in sufficient quantities at this site in Ostia for us to assume that they were widely or extensively exported. (46) But the Baths of the Swimmer is only one site in Ostia; it may be that more Lusitanian amphorae will turn up in excavations in other parts of the port.

The distribution of Lusitanian Garum I, II, III and IV amphorae is shown in Figures 6.8, 6.9, 6.10 and 6.11. The significance of such distribution maps of artefacts is somewhat limited, but it does help to give some sort of indication of the general horizons of the export of Lusitanian surplus.

FIGURE 6.7 LUSITANIAN GARUM IV AMPHORAE

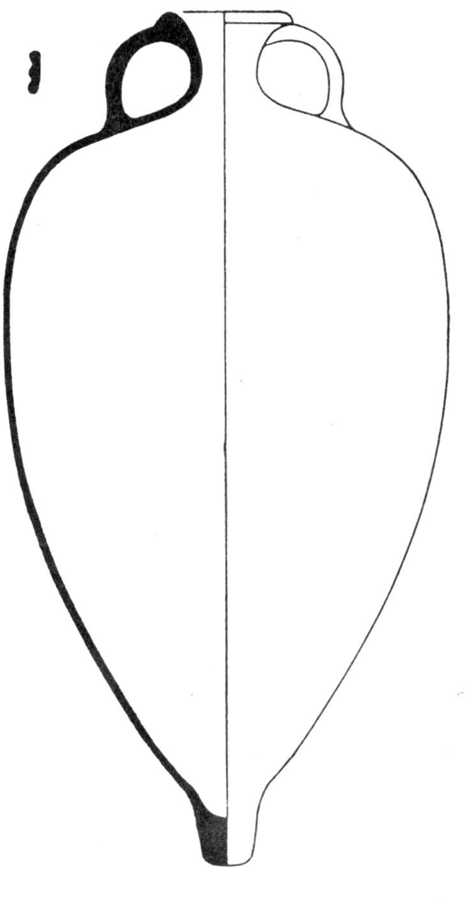

(after G. Cardoso, Conimbriga, 17, 1978, fig. IX, no. 20; scale 1:6)

FIGURE 6.8 DISTRIBUTION OF LUSITANIAN GARUM I AMPHORAE

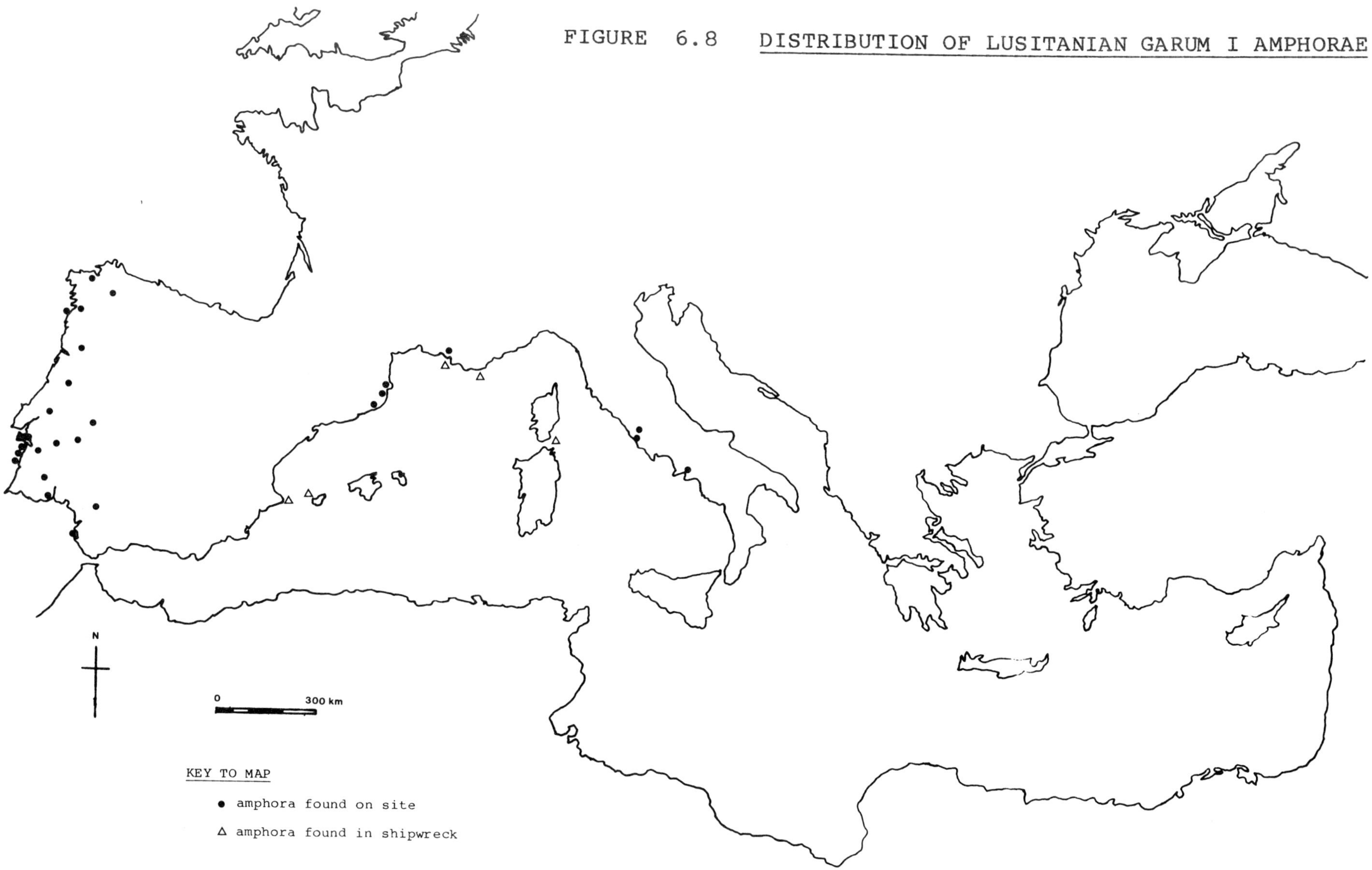

FIGURE 6.9 DISTRIBUTION OF LUSITANIAN GARUM II AMPHORAE

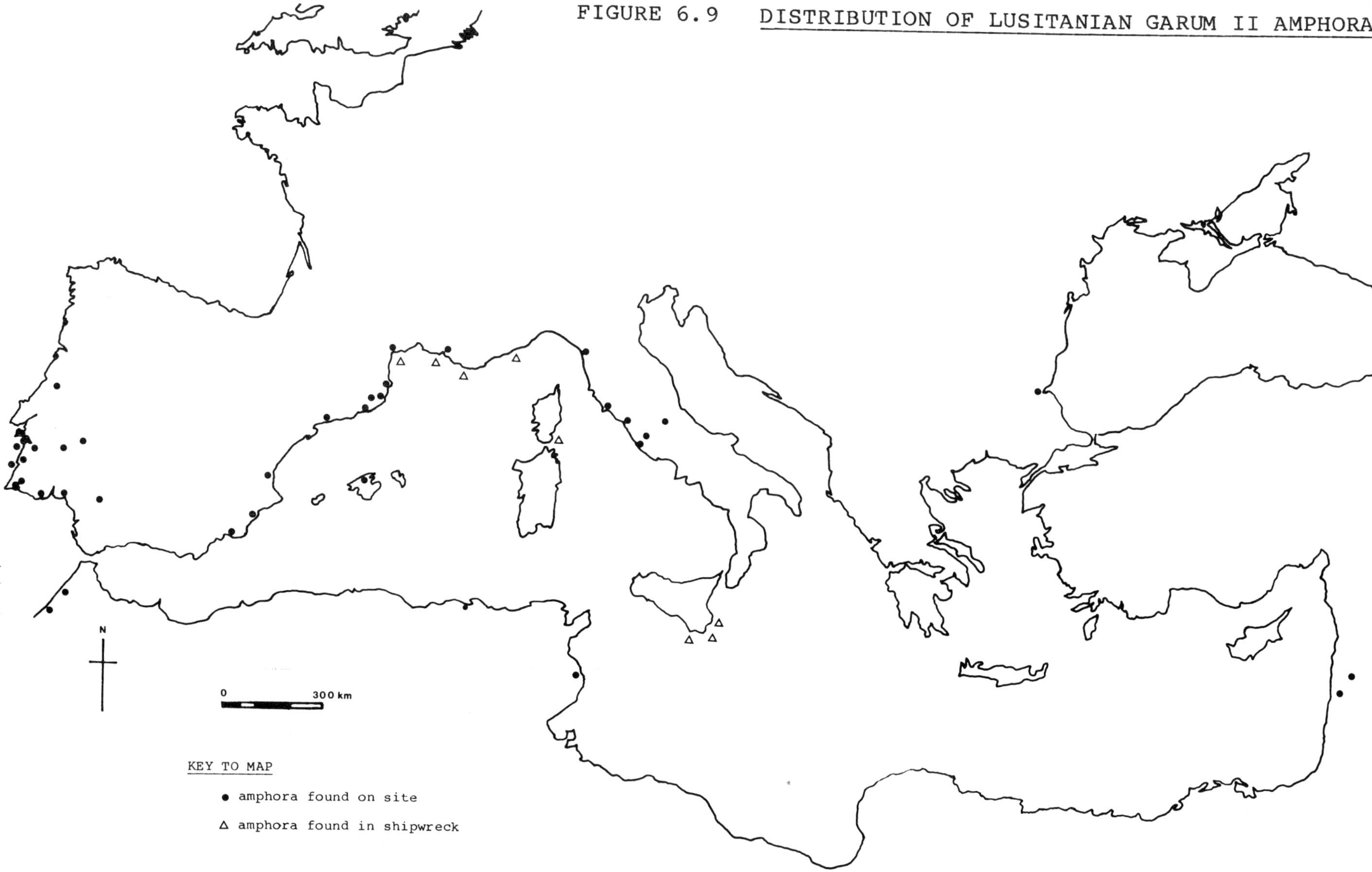

FIGURE 6.10 DISTRIBUTION OF LUSITANIAN GARUM III AMPHORAE

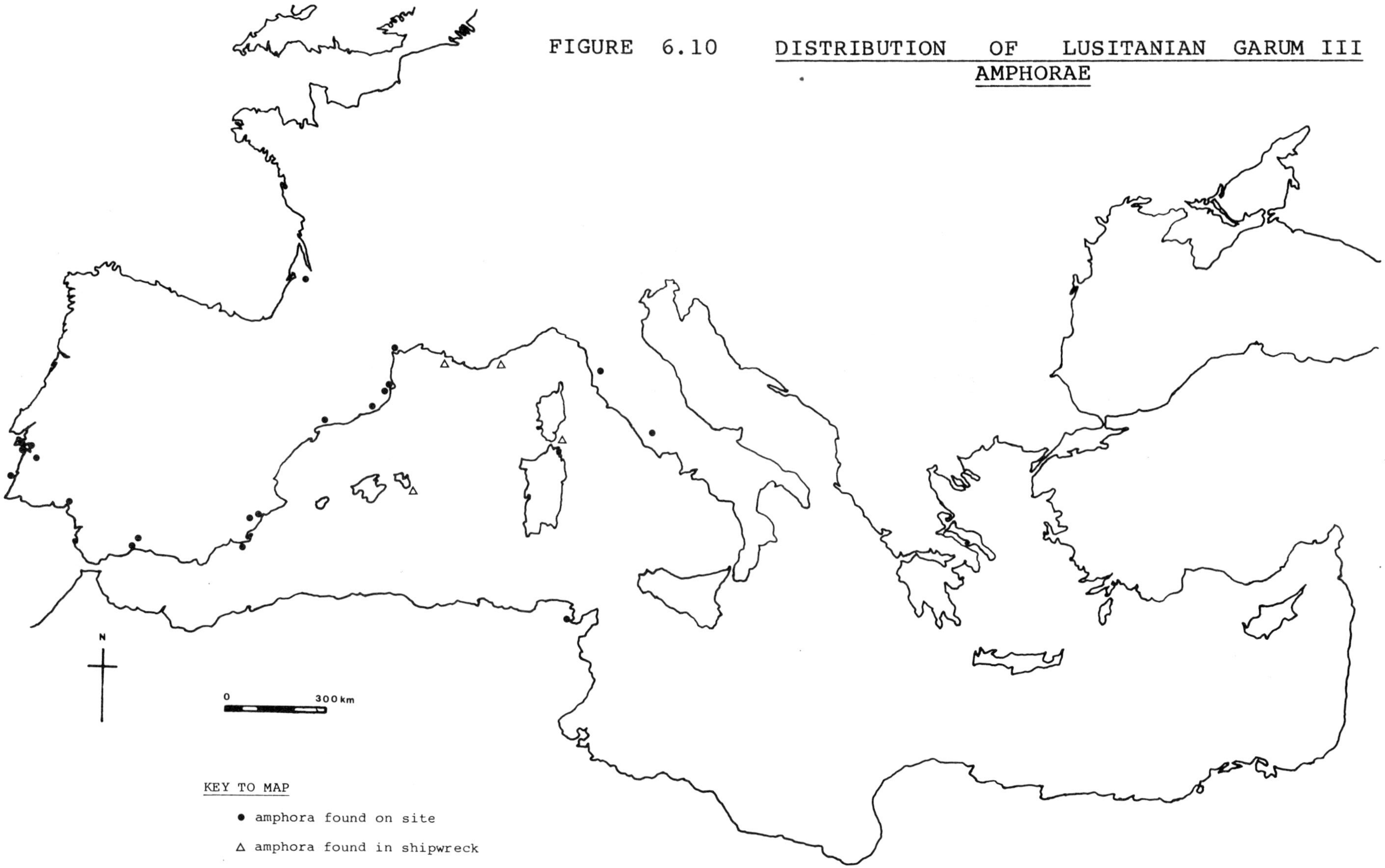

FIGURE 6.11 DISTRIBUTION OF LUSITANIAN GARUM IV AMPHORAE

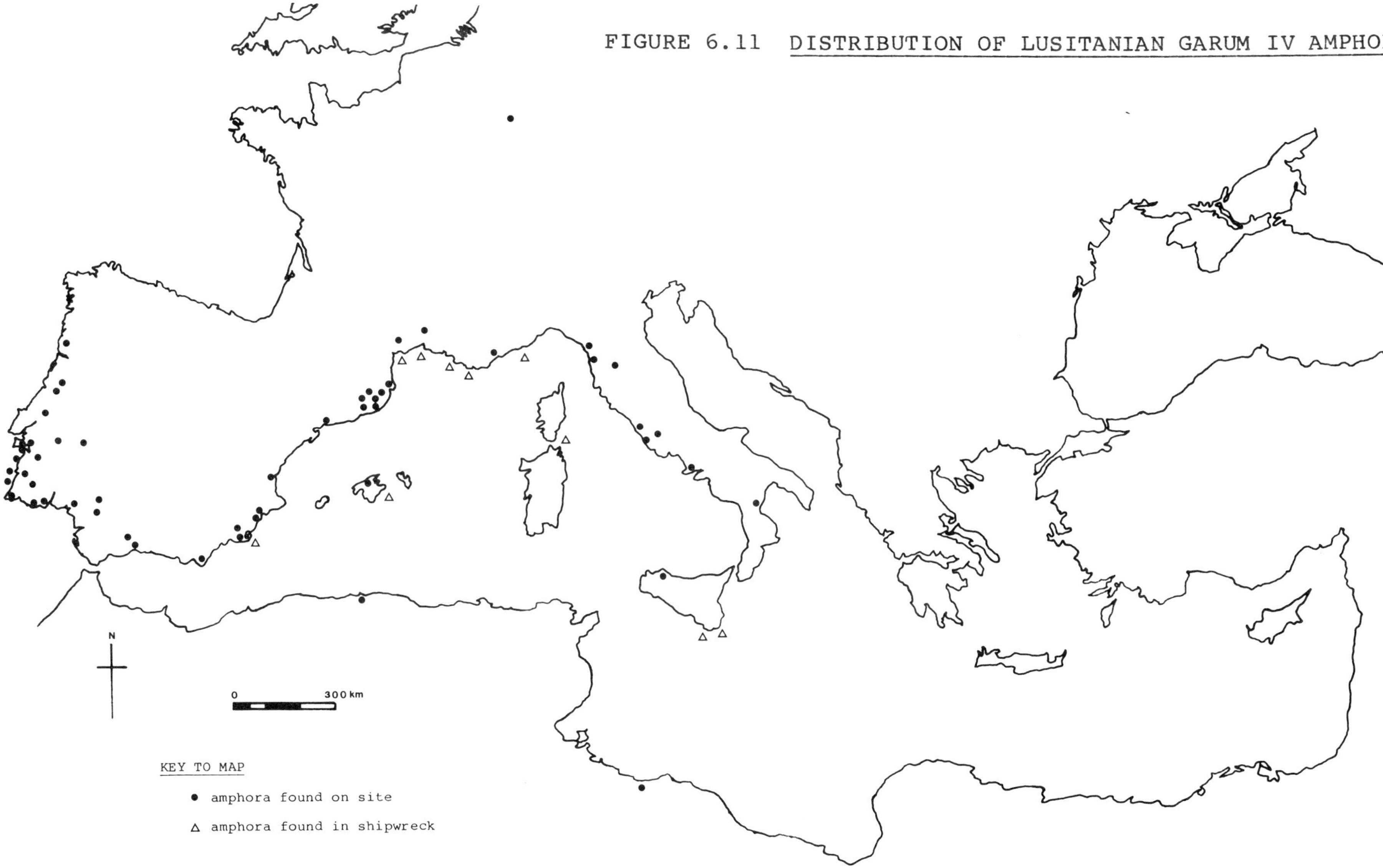

The distribution of Lusitanian Garum I is in general limited to the Iberian peninsula (essentially Lusitania) and Ostia/Rome. The other findspots all suggest that the products were en route for Rome. It is impossible, however, from this distribution of amphorae to quantify the level of the export of Lusitanian fish-products. But of the wrecks containing such amphorae, those off Conejera Island, Ibiza and Cap Bénat, Var seem to have been carrying cargoes in large part made up of these Lusitanian products. (47)

A broadly similar pattern is revealed for the later types, Lusitanian Garum II, III and IV; their distribution seems also to have been generally restricted to Lusitania and the sea route to Rome via Ostia. From this it seems that the horizons of the Lusitanian fish-processing industry had not significantly changed from the mid-first to the late-fourth centuries A.D.; there had been no dramatic expansion or contraction in the markets where this produce was exchanged.

However, just occasionally examples have been found off the route to Rome via Ostia : so examples of Lusitanian Garum II type have been found at Bourgas on the Black Sea and at Sebaste and Samaria in Palestine, while examples of Lusitanian Garum IV have been found on the north African coast at Caesarea and Sabratha, in the Canary islands at Arrecife and at Augusta Treverorum right in the interior of Gallia Belgica. This last example may well reflect the trade route up the Rhone and the Rhine, so vital for keeping the Roman army supplied with food and equipment on the frontier. (48)

Thus, there seem to have been three main mechanisms for the exchange of Lusitanian surplus fish-products. First, the local market. As Hopkins has recently observed, since the cost of long-distance transport by land or sea was a major constraining factor on all but the very rich, most Roman producers would have disposed of their surplus produce on the local market. (49) An obvious new market - after the institution of the province - was the provincial capital, Augusta Emerita, with its large non-productive population of consumers, from where examples of both Lusitanian Garum II and especially Lusitanian Garum IV amphorae have been found. But why, one wonders, as yet no Early Empire types - for example, Lusitanian Garum I ?

The amphorae thus far discussed were the only Lusitanian types that are found a long distance from their area of production, but there were many local varieties whose diffusion seems to have been limited to the province itself. For example, a variant of the Beltran IIb type was produced in a kiln near Lagos, where a large dump of such amphorae has been found. This type has its handles joined onto the outer edge of its lip, a stylistic feature observable in the Lusitanian Garum II type. (50) Another local type of very limited diffusion was also produced in the Algarve at the kiln of São João da Venda. (51) All these amphorae were probably produced on a rural estate to be filled with the produce of that estate, a practice attested in the Digest and suggested by those amphora stamps which bore the name of an estate owner. (52)

It is interesting that these more local types were produced in the Algarve rather than in the Sado estuary; it would seem to confirm conclusions drawn earlier about the organization of production. For in the Algarve units of production were more often smaller, and more isolated and independent one from another than in the Sado estuary; thus individual concerns in the Algarve, for the most part located on agricultural estates, would have produced their own amphorae, to transport surplus produce. This would explain the variety of local types of amphorae produced. In the Sado estuary there was a much greater concentration of units of production and hence a greater homogeneity in the types of amphorae used.

The second mechanism of exchange was long-distance, maritime transfer of surplus along the route to Rome via Ostia. But the amphorae found off this route to Rome are economically more interesting than those found on it, and provide the evidence for the third mechanism of exchange, namely inter-provincial or inter-regional transfer. For the maritime route to Rome represented the regular, established link between periphery and imperial centre. It owed its existence, and its importance to the regular flow of taxes to the imperial capital and the need to keep Rome, with its enormous population of mostly non-productive consumers, supplied with vital materials, essentially food. It was what might be termed the "official" trade route, and as such the obvious first choice for anyone who had a surplus to export. But the transfer of produce between provinces or regions presupposes that there was an even greater surplus to be disposed of on markets not so regularly and traditionally linked, and only considered after the obvious market - Rome - had been supplied. (53)

The development of north African exports from the mid-third century A.D. onwards has received much scholarly attention of late; it is illustrative of the way in which the export of Lusitanian products may have developed. To start with, surplus agricultural goods and fineware pottery (African red-slip ware, terra sigillata clara) were exported only to Rome; but as the level of surplus increased in the later Empire, African products started to satisfy the demands of a much wider and varied market. (54)

To obtain more precise information on the distribution of Lusitanian garum and other fish products, it is possible to consider again the distribution of the stamped amphorae (see Figure 6.7). Unfortunately, as we have seen, only the Lusitanian Garum II type amphora was regularly stamped, although some samples of stamped Lusitanian Garum I amphorae are starting to emerge from the Sado estuary. (55) It is now accepted that such amphora stamps record the name of the owner of the estate on which the vessels were produced. (56) They do not necessarily reveal the identity of any producers of fish preserves, still less the shippers of such products. But if, as I have argued, the production of fish-sauces was in many instances controlled by larger landowners, then it is likely that the estate owner would have produced these amphorae to transport his own surplus garum, or that of his closer neighbours.

Only four particular stamps are at all useful : LEV.GEN., AEMHEL., OLYNT. and Q.E.TAVR. The general pattern of the distribution of these is in broad terms very similar to that of the unstamped amphorae themselves; it is restricted to the production zones of Lusitania, neighbouring areas of the Iberian peninsula, southern Gaul and Italy - in other words, the trade route that linked Lusitania to the imperial capital. The isolated find of the stamp LEV.GEN. in north Africa is also consistent with the scatter of the unstamped amphorae, and represents what must have been a common, short-distance, secondary exchange route - between southern Iberia and north Africa.

Transport costs were such that it was only the very rich, with a very large surplus to dispose of, who would have engaged in inter-provincial and long-distance trade. (57) This is borne out by an examination of the composition of the entire cargoes of wrecks, in which Lusitanian fish-products have been discovered (see Table 6.12).

TABLE 6.12

THE COMPOSITION OF CARGOES OF

SHIPWRECKS CONTAINING LUSITANIAN AMPHORAE

A. WRECKS CONTAINING LUSITANIAN GARUM I

1. SAINT GERVAIS (Fos-sur-Mer)

Lusitanian Garum I
Dressel 20 (Baetica, oil)
Pelichet 47 (Gaul, wine)
Dressel 7-11 (Baetica, *garum*)
Dressel 28 (*defrutum*)
Late Rhodian (wine)

Ref. : *Archaeonautica*, 2, 1978, 109-81

2. CONEJERA ISLAND (Ibiza)

Lusitanian Garum I (stacked in 3 layers in 6 files)
1 other unspecified form of amphora
? Lead ingots

Refs. : *NAH Arq.*, 6, 1962, 177-88
& 7, 1963, 188-94

3. CAP BÉNAT I (Var) Date : 1st half Cl A.D.

Homogeneous cargo of Lusitanian Garum I amphorae
1 x Ibizan amphora

Ref. : *Cah. d'arch. sub.*, 2, 1973, 137-145;
Archaeonautica, 5, 1985, 152 & fig. 118

4. ANSE DE LA FONTAINE
(Javea, Majorca)

Lusitanian Garum I

Ref. : _Cah. d'arch. sub._, 5, 1976, 99-103

5. SUD-LAVEZZI 3 (Corsica) Date : 1st half C1 A.D.

Lusitanian Garum I (1)
Dressel 2-4 (Catalonia, wine) (100)
Pascual 1 (Tarraconensis, wine) (1)

Ref. : _Gallia_, 40, 1982, 446-50

B. <u>WRECKS CONTAINING LUSITANIAN GARUM II</u>

1. ANSE GERBAL I (Port-Vendres)

Lusitanian Garum II
Lusitanian Garum IV
Cylindrical (Africa, oil)
Small African amphorae
African _terra sigillata_
Glassware

Refs. : _Gallia_, 17, 1959, 450-1 & 22, 1964, 475;
RStLig., 37, 1971, 10

2. CAPE OGNINA (Syracuse)

Lusitanian Garum II
Africana I (Africa, oil) (90%)
Kapitan I & II (Africa, oil)
Egyptian blue dye
African coarseware
Glassware

Refs. : _IJNA_, 3.1, 1974, 150-3;
MEFRA, 93, 1981, 376

3. MARZAMEMI F (Sicily)

Lusitanian Garum II
Lusitanian Garum IV
Africana IIc (Africa, oil)

Refs. : <u>IJNA</u>, 5.4, 1976, 347;
<u>Ostia IV</u>, 277-278
<u>IJNA</u>, 10.4, 1981, 328

4. SUD-LAVEZZI I (Corsica)

Lusitanian Garum II (83 of c. 450)
Lusitanian Garum III (194)
Flat-bottomed amphorae : garum jars ? (113)
Lusitanian Garum IV (6)
Dressel 23 (Africa, oil) (3)
Small piriform cf. Beltran 72 (unspecified)

Refs. : <u>Gallia</u>, 40, 1982, 437-44
<u>IJNA</u>, 5.4, 1976, 347

5. POMMEGUES (Marseilles)

Lusitanian Garum II
12 other types (not all identified), including :
Pelichet 47 (Gaul, wine)
African <u>terra sigillata clara</u>

Ref. : <u>Archaeonautica</u>, 2, 1978, 101-7

6. PLANIER 7 (Marseilles)

Lusitanian Garum II
Lusitanian Garum IV
African cylindrical (Africa, oil)

Ref. : <u>Gallia</u>, 20, 1962, 157-9, 161 (fig. 26)

C. WRECKS CONTAINING LUSITANIAN GARUM III

1. FAVARITZ, CALES COVES
   (Menorca)

   Ref. : M. Fernandez-Miranda & M. Belen, El Fondeadero de Cales Coves (= EAE, 101), Madrid, 1979

2. SUD-LAVEZZI I (Corsica)

   See under B.4

3. CATALANS (Marseilles)

   Lusitanian Garum III (2)
   Type cf. Dressel 20 (? oil)
   Cylindrical (Africa, oil)
   Small amphora

   Ref. : Gallia, 31, 1973, 585-6

4. DRAMONT F, SAINT RAPHAEL

   Lusitania Garum III (1 of c. 120)
   Cylindrical (Africa, oil) (85)
   unspecified amphorae (c. 35)

   Ref. : Gallia, 33, 1975, 600-1 & fig. 38

D. WRECKS CONTAINING LUSITANIAN GARUM IV

1. ANSE GERBAL I (Port-Vendres)

   See under B.1

2. MARZAMEMI F (Sicily)

   See under B.3

3. LAVEZZI (Corsica)

See under B.4

4. CAP BLANC (Majorca)

Lusitanian Garum IV
African amphorae

Refs. : IJNA, 5.4, 1976, 347;
Ostia IV, 278

5. FEMINA MORTA (Ragusa, Sicily) Date : C4 A.D.

Lusitanian Garum IV
Africana I, IIc & IId
"Spatheia" type amphorae (Africa)
African *terra sigillata*

Refs. : IJNA, 4.2, 1975, 385;
Kokalos, 22-3, 1976-7, 627, Tav. CXXXV, 10;
MEFRA, 93, 1981, 374

6. BAY OF PAMPELONE (Var)

Lusitanian Garum IV
Cylindrical (Africa, oil)
Africana Grande (Africa, oil)
2 types (Mauretania)
Beltran 72 (S.Spain, *garum* ?)

Refs. : RAN, 9, 1976, 185-6;
MEFRA, 90, 1978, 382

7. CHRÉTIENNE D (Agay-Anthéor)

Lusitanian Garum IV

Ref. : RAN, 9, 1976, 186-7 & fig. 9

8. PLANIER 7 (Marseilles)

See under B.6

There are again problems over the evidence, since it is impossible to be sure if the entire ancient cargo has been recovered. Often part of the cargo (especially precious metals) was salvaged either at the time of the shipwreck itself or subsequently. Moreover, reports of underwater excavations are often afflicted by a lack of quantification, so that it is difficult to assess the relative proportions of the various goods which made up the ancient cargo. What is generally clear is that, to date, few wrecks have been recovered with distinctly Lusitanian products as part of their cargoes, especially from the early Empire. Lusitanian Garum I amphorae have only yet been found on five wrecks. This very fact might signify that Lusitanian products were not widely exported in this period. But such a conclusion would be dangerous for the following reasons. First, it is not yet firmly established to what extent Lusitanian products were shipped in "Baetican" amphorae; secondly, barrels might also have been used for the shipment of these products and thirdly, it may just reflect the chance non-survival of wrecks in which Lusitanian products were shipped. However, two of these five wrecks do seem to have been carrying cargoes made up almost entirely of Lusitanian fish-products : the wrecks at Conejera Island and Cap Bénat I.

It must have been rare for a single producer to have a sufficient surplus to fill an entire cargo, although the non-perishable nature of these fish products would have allowed him to store them until he did have enough. More significantly, it also allowed him to store them until there was a shortage of supply, perhaps caused by a bad fishing season, and hence an increased demand for such products. He would then be able to sell both luxury and more commonplace grades of his product at prices higher than normal. Many ancient writers comment caustically on the high price of the best _garum_; this most likely reflects the scarcity, and variability in supply, of the best tunny needed for the best _garum_, but it could also imply that some large-scale producers hoarded their product to ensure a good price for it. (58)

Smaller producers would probably have needed to dispose of their surplus as quickly as possible and so sold it to shippers to form only a small part of their cargoes. The presence of some possibly Lusitanian tin ingots in the cargo of the Port-Vendres II wreck containing predominantly Baetican olive oil and fish products perhaps reflects a mechanism whereby some Lusitanians may have sold their surplus to shippers from neighbouring provinces. (59)

The cargoes which contained Lusitanian Garum II, III and IV amphorae comprise a great variety of goods from many different origins. There does, however, seem to have been a tendency to ship Lusitanian products together with African foodstuffs and fineware pottery. For example, the cargo of the Anse Gerbal I wreck contains mostly Lusitanian amphorae, but with some African cylindrical and small amphorae and African terra sigillata; but, more dramatically, the Sud-Lavezzi 1 wreck carried only a small cargo of some four hundred and fifty amphorae, but a large proportion of these comprised three types of Lusitanian amphorae. It remains a possibility that the one hundred and thirteen flat-bottomed "amphorae" reported from this last wreck were also carrying fish products, since their description matches in most respects that of the garum pots familiar from Aemilius Scaurus' garum shop in Pompeii. (60). However, in most other wrecks the proportion of Lusitanian to African goods is reversed : so in the Cape Ognina wreck Africana I amphorae made up some ninety per cent. of the cargo and in the wreck from Pampélone (Var) Baetican, Tripolitanian, Mauretanian and Lusitanian goods were all shipped together. (61)

The overall conclusion to be drawn from this evidence seems to be that Lusitanian products rarely made up an entire cargo at any period, but were usually shipped with products from other provinces (essentially Baetica in the early Empire and Africa in the later period). A passage of the Digest makes it clear that it was a common practice for number of merchants to combine a variety of goods from a variety of production areas into a single cargo for shipment across the Mediterranean. (62) The volume of the Lusitanian surplus was not great enough to encourage the regular shipment of whole cargoes from the Atlantic to the Mediterranean. Instead, the producers of Lusitania with a surplus to export operated through the shippers of Baetica and, later, north Africa.

Baetica in the early Empire and north Africa in the later Empire were producing large agricultural surpluses (especially of olive oil, and in the case of north Africa grain as well), which were shipped regularly to Rome. The fact that Lusitania was relying on the structures of maritime transport established for the basic use of these other western provinces may also in part explain the small trace that Lusitanian traders and shippers have left on the historical record. Lusitanian products would have had to have been transferred to the ports of Baetica or north Africa, before they began their journey eastwards. The *codicarii* may have helped to effect this transfer. Alternatively, Baetican and/or north African shippers could have come themselves to the production areas to acquire Lusitanian *garum*, either to supplement fish products purchased in their own provinces, or to make up for an unforeseen shortfall in production in their own areas. Numerous Baetican and north African oil amphorae have come to light in the Sado estuary, which perhaps bear witness to this kind of exchange. (63)

iii. Conclusion

The study of Lusitanian garum amphorae has given some indication of the overall importance of the fish-processing industry to the Lusitanian economy. Since there is no secure evidence for the export of wine or olive-oil, fish-products and metals were probably the two major resources of the province exported overseas on any sort of scale. The study of the amphorae allows some general conclusions to be drawn, first, about the chronological development of the industry and, secondly, about the general distribution of Lusitanian commodities over the entire Roman Empire.

To summarize, the industry in Lusitania began very much as an adjunct to that of Baetica, centred on Gades. Surplus Lusitanian produce was distributed in "Baetican" forms of amphora in the early years of the Principate and so it is extremely difficult to assess the independent contribution of Lusitania. It was only in the mid-first century A.D. that characteristically Lusitanian forms of amphora began to be produced, which symbolise the emergence of an independent Lusitanian branch of the industry. These Lusitanian forms of amphora were to gain some ascendancy over the earlier Baetican forms in certain well-investigated centres (for example, at the Baths of the Swimmer, Ostia). Whether this, as Parker suggested, indicates that there was a shift in production from Baetica to Lusitania is uncertain. (64) The date of the appearance of a distinctly Lusitanian form of amphora is not surprising, since it would have taken some time for Lusitania to become fully integrated within the economic system of the Roman Empire; the mechanisms for the exchange of surplus produce could not appear overnight.

The geographical distribution of the later forms of Lusitanian amphora (Lusitanian Garum II, III and IV) reveals that the horizons of the Lusitanian garum industry were, in large part, the same in the later Empire as they had been during the early Principate; in no way does it suggest that there was a significant decline in long-distance exchange from the third century onwards. (65) All the evidence points to an industry functioning in a similar way, and possibly also on a similar scale, to that of the early Empire. It is, however, impossible to quantify the evidence in such a way that any assessment of the scale of production can be made. A mere count of surviving amphorae is methodologically unsatisfactory.

Evidence is also lacking for Lusitanian shippers and traders. It would be dangerous from this very fact to argue from silence that Lusitanian trade was scant compard with, say, Baetica, for whose traders there is abundant evidence. For much of our information about Baetican traders comes from the dump of olive-oil amphorae from the Monte Testaccio in Rome. If a commensurate dump of Lusitanian fish-sauce amphorae were discovered, then our knowledge of Lusitanian traders would vastly increase. However, the configuration of cargoes from wrecks carrying Lusitanian products (admittedly small in number) suggests that Lusitanian goods were only rarely carried as homogeneous cargoes. They were much more likely to form part of a cargo of essentially Baetican, or later north African, goods. From this it follows that a consistently large surplus of fish preserves was not being produced in Lusitania to warrant regular export. The surplus seems mostly to have been distributed within the province of Lusitania or to have been exported to the market at Rome via Ostia. Most of the finds of Lusitanian amphorae can be explained in terms of this trade route. The finds of, first, Baetican and, later, north African oil amphorae in the Sado estuary, the major production area for Lusitanian fish products, provide further clues as to the mechanisms of exchange. Baetican and African traders perhaps exchanged this oil for Lusitanian garum, with which they filled part of their cargoes.

The internal evidence from the production centres suggests that the industry operated at two separate levels, just as mining has been shown to have done. For fish-processing ranged from being an subsidiary, seasonal activity on estates where agriculture was the main concern, to being the central activity of a large, nucleated industrial complex, such as at Tróia. Even at Tróia the operations may also have ranged from the small- to the large-scale, as occurred elsewhere in Roman industry - for example, at La Graufesenque. The layout of Tróia suggests that the work was carried out by small units of men of relatively low social status, sometimes formed into associations for mutual benefit, sometimes under a contractural relationship to a wealthy individual, who could have provided the necessary input of capital to ensure the successful running of the diversified activities involved in fish-processing. Processing also took place in some cities, at Setúbal (= Roman Caetobriga ?), Olisipo and Ossonoba.

NOTES : CHAPTER SIX

1. C. Amaro, 'Casa dos Bicos - noticia historico-arqueologico', Arqueologia (Porto), 6, dezembro 1982, 96-111, esp. 99; traces of Roman quay in modern Convento de S. Domingos (Rossio & Rua das Canastras): I. Moita, 'Problemas de Lisboa romana. A recuperação do teatro de Olisipo' in Arqueologia de las Ciudádes modernas superpuestas a las antiguas (Zaragoza, 1983), Madrid, 1985, 285-302, esp. 288.

2. See above p. 39 and note 49.

3. IRCP 35 = AE 1981, 490. A. Alarcão, C. Beloto, J. d'Encarnação, M.M. de Almeida, 'O mosaico do Oceano de Faro', AMF, 10, 1980, 219-32; J. Lancha, 'La mosaique d'océan découverte à Faro (Algarve)', Conimbriga, 24, 1985, 151-75.

4. J.M. Alvarez Martinez, El puente romano de Mérida (Monografias Emeritenses, 1), Badajoz, 1983, 10-11; J. de Alarcão, Portugal romano, Lisbon, 1983, 51.

5. J. Soares, 'Sado : navegação, pesca e comercio. Algumas notas historicas' in H. Cabeçadas, J. Soares & A.J. Guerra (ed.), Embarcações tradicionais do Sado, Setúbal, 1986, 7-21, esp. 14 & fig. 7 (photo). For a possible port discovered recently on the right bank of the river Mira at Vila Nova de Milfontes, Odemira (Beja) see Informação arq., 5, 1982-83, 9.

6. On Spanish traders in general, T.R.S. Broughton, 'Some notes on trade and traders in Roman Spain' in J.A.S. Evans (ed.), Polis and Imperium : studies in honour of E.T. Salmon, Toronto, 1974, 11-30. For some Lusitanian traders : M. Beltran Lloris, Las anforas romanas en España, Zaragoza, 1970, navicularii nos. 10 (p. 219), 92 (p. 233), 122, 125 (p. 238), ?145 (p. 241), but note his reservations (p. 264); for more confidence, A.J. Parker, 'Lusitanian amphoras' in Méthodes classiques et méthodes formelles dans l'étude des amphores (=

Collection de l'école française de Rome, 32), Paris/Rome, 1977, 35-46, esp. 35. E. Rodriguez Almeida, 'Novedades de epigrafia anforaria del Monte Testaccio' in Recherches sur les amphores romaines (= Collection de l'école française de Rome, 10), Paris/Rome, 1972, 107-241 seems to prove that Beltran's nos. 10, 92, 122 were all Baetican navicularii (from Astigi and Hispalis).

7. CIL VI 9677 = ILS 7278; CIL VI 9873

8. CIL XIII 1966

9. R.I. Curtis, 'Negotiatores allecarii and the herring', Phoenix, 38, 1984, 147-58.

10. CIL VI 9676

11. L. Casson, 'Harbour and river boats of ancient Rome', JRS, 55, 1965, 31-39, esp. 39.

12. CIL II 260; V. Gil da Cruz Soares Montas, 'Notas acerca tres inscrições de Olisipo', Conimbriga, 15, 1976, 158-161.

13. CIL II 25

14. ILS 687 : codicarii nabiculari/infernates; ILS 1433 : codicarii, navicularii et quin. corp. navigantes

15. On the size of merchantmen : A. Tchernia & P. Pomey, 'Le tonnage maximum des navires de commerce romaine', Archaeonautica, 2, 1978, 233-251; on the ship-owners : K. Hopkins, 'Models, ships and staples' in P. Garnsey & C.R. Whittaker, Trade and Famine in Classical Antiquity (= Cambridge Philological Society, Suppl. Vol. 8), Cambridge, 1983, 84-109, esp. 96-102. In general J.H. d'Arms, Commerce and Social Standing in Ancient Rome, Cambridge (Mass.), 1981.

16. P. Olitius Apollinaris : CIL XII 4406, CIL XV 3974, 3975; Sex. Fadius Secundus Musa : CIL XII 4393 = ILS 7259, CIL XV 3863-3873; Q. Fulvius Carisianus : CIL II 1064 = ILS 6919, CIL XV 3876. In general, see A. Tchernia, 'D. Caecilius Hospitalis et M. Iulius Hermesianus (CIL VI 1625b & 20742)' in J.M. Blazquez (ed.), Producción y comercio del aceite en la antigüedad, I, Madrid, 1981, 155-160 and J.H. d'Arms, op.cit. (n. 15). P. Garnsey, 'Grain for Rome' in P. Garnsey, K. Hopkins & C.R. Whittaker, Trade in the Ancient Economy, London, 1983, 118-30, esp. 125-6.

17. 3.5.3 (168); cf. the wreck discovered at Sancti Petri (Cadiz) carrying copper and lead ingots from the early C1 A.D. : O. Vallespin, 'Carta arqueologica de La Caleta' VI Congreso Internacional de Arqueologia Subacuática (Cartagena, 1982), Madrid, 1985, 59-74, esp. 69; id., 'The Copper Wreck (Pecio del Cobre)', IJNA, 15.4, 1986, 305-322.

18. M. Maia, 'Cepos de chumbo de ancoras romanas encontrados ao largo de Sesimbra', Setúbal Arq., 1, 1975, 177-80; J. Soares, art. cit. (n. 5).

19. P. Diaz Alvarez, Anforas romanas en el eje atlantico galaeco-lusitano, Vigo, 1984; J. Naveiro Lopez, 'As anforas romanas de A Coruña (I)', Brigantium, 2, 1981, 117-26; id. 'As ánforas romanas de A Coruña (II)', Brigantium, 3, 1982, 63-74; X. Gonzalez Fernandez, 'O comercio romano na ria de Vigo', Brigantium, 4, 1983, 73-98; F.S. Lopez, 'La subacuática en los costas del norte y noroeste peninsular : estado de la cuestión', VI Congreso Internacional de Arqueologia Subacuática (Cartagena, 1982), Madrid, 1985, 33-58.

20. C. Sevius Lupus Conimbrigensis : CIL II 2559; on the lighthouse see S. Hutter, Der Römische Leuchtturm von La Coruña (Madrider Beitrage, 3), Mainz-am-Rhein, 1973; T. Hauschild, 'Der römische Leuchtturm von La Coruña (Torre de Hercules). Probleme seiner Rekonstruktion', MDAI(M), 17, 1976, 233-258.

21. So a ship of 500 tonnes could carry 2600 hectolitres of wine in amphorae or 4600 hectolitres in barrels :

A. Tchernia & P. Pomey, art.cit. (n. 15), esp. 250.

22. F. Braudel, The Structures of Everyday Life : the limits of the possible (= Civilization and Capitalism, 15th - 18th Century, I), London, 1981, 215.

23. See F. Zevi, 'Appunti sulle anfore romane', Archeologia classica, 18, 1966, 229-247 for an updating of Dressel; for the difficulties of typology : J.-P. Morel, 'La céramique comme indice du commerce antique (réalités et interpretations)' in P. Garnsey & C.R. Whittaker, op. cit.(n. 15), 66-74; for greater optimism, G. Pucci, 'Pottery and trade in the Roman period' in P. Garnsey, K. Hopkins & C.R.Whittaker (ed.), op. cit. (n. 16), 105-117.

24. D.P.S. Peacock, 'Amphorae and the Baetican fish industry', Antiq. Journal, 54, 1974, 232-243, esp. 232.

25. D.P.S. Peacock, 'Roman amphorae : typology, fabric and origins' in Méthodes classiques et méthodes formelles dans l'étude des amphores, Paris/Rome, 1977, 261-278.

26. H. Cockle, 'Pottery manufacture in Roman Egypt', JRS, 71, 1981, 87-97.

27. On these kilns, see A. Coelho-Soares & C. Tavares da Silva, 'Anforas romanas da Quinta da Alegria (Setúbal)', Setúbal Arqueologica, 5, 1979, 205-221 and 'Anforas romanas da area urbana de Setúbal', Setúbal Arq., 4, 1978, 171-201; G. Cardoso, 'Anforas romanas no Museu do Mar (Cascais)', Conimbriga, 17, 1978, 63-78; A.M. Dias Diogo, 'Fornos de anforas do Monte do Bugio. Noticia preliminar', Conimbriga, 19, 1980, 147-150.

28. M. Maia, 'As anforas de S. Bartolomeu de Castro Marim', Clio, 1, 1979, 141-148; J. Leite de Vasconcelos, 'Olaria luso-romana em S.Bartolomeu de Castro Marim' , AP, 4, 1898, 329-336 and 'Da Lusitania a Betica', AP, 5, 1900, 225-249.

29. G. Cardoso, art.cit. (n. 27), 69 and Catalogue no. 11.

30. B. Liou & R. Marichal, 'Les inscriptions peintes sur amphores de l'anse Saint-Gervais à Fos-sur-Mer', Archaeonautica, 2, 1978, 109-81, esp. 141, no. 12 & fig. 16.

31. A. J. Parker, art. cit. (n. 6), 38.

32. For references for amphora finds see Appendix VI.

33. C. Tavares da Silva et al., 'Escavações arqueologicas no Castelo de Alcácer do Sal (campãnha de 1979)', Setúbal Arq., 6-7, 1980-81, 149-218, esp. 197-200.

34. A. J. Parker, art. cit. (n. 6), 39; on Baetican garum we await with anticipation : M. Ponsich, Factores geo-economicos de Bética meridional : el aceite de oliva y las salazones de pescado, Madrid (forthcoming).

35. For example, note the remains of ? Sardina pilchardus in an amphora from the Anse Gerbal wreck at Port-Vendres : Y. Chevalier & C. Santamaria, 'L'épave de l'anse Gerbal à Port-Vendres', Omaggio a F. Benoit, Bordighera, 1971, V, 7-32.

36. S.J. Keay, Late Roman Amphorae in the Western Mediterranean : a typology and economic study : the Catalan evidence, 2 vols. (B.A.R. International Series), 196, Oxford, 1984.

37. A.J. Parker, art.cit. (n. 6), 36-37.

38. G. Cardoso, art.cit. (n. 27), 71.

39. V. Gonçalves, H. Catarino, A.M. Arruda, 'O sitio romano-arabe do Vale de Boto', Clio, 2, 1980, 71-9, esp. 72-5.

40. S.J. Keay, op. cit. (n. 36), 151, 155.

41. M. L. Veiga Silva Pereira, 'Marcas de oleiros algarvios do periodo romano', AP, iii, 7-9, 1974-1977, 243-268.

42. C. Tavares da Silva, J. Soares & L. Ferrer Dias, 'Trabalhos arqueologicos na Ilha do Pessegueiro (1980)', Setúbal Arq., 6-7, 1980-81, 219-47; C. Tavares da Silva, 'Escavações arqueologicas na ilha do Pessegueiro', Almadan, 2, 1983-84, 20-22; C. Tavares da Silva et al., 'Escavações arqueologicas na Ilha do Pessegueiro (Sines) : noticia de 2a. campanha (1981)', Arq. de Beja, ii, 1, 1984, 11-45.

43. S.J. Keay, op. cit. (n. 36), 156-168.

44. ibid., 645.

45. A. Coelho-Soares & C. Tavares da Silva, art. cit. 1 (n. 27), 211, 214.

46. A. Carandini & C. Panella, 'The trading connexions of Rome and central Italy in the late second and third centuries : the evidence of the Terme del Nuotatore excavations, Ostia' in A. King & M. Henig (ed.), The Roman West in the Third Century : contributions from archaeology and history (BAR Int. Ser., 109), Oxford, 1981, II, 487-503, esp. 498.

47. See below and especially Table 6.12.

48. On which see now J. Remesal Rodriguez, La annona militaris y la exportación de aceite betico a Germania, Madrid, 1986.

49. K. Hopkins, art. cit. (n. 15).

50. A.J. Parker, art.cit. (n. 6), 39.

51. A. dos Santos Rocha, 'Noticia de algumas estações romanas e arabes do Algarve', AP, 1, 1895, 193-212; M.L. Estacio da Veiga Affonso dos Santos, *Arqueologia romana do Algarve*, II, Lisbon, 1972, 166-170.

52. *Dig.* 33.7.25.1 and 8.3.6.pr.; stamps : FIGUL.CEPAR. (M.H. Callender, *Roman Amphorae : with index of stamps*, London, 1965, 97, no. 303; EX F(iglinis) DECENTI (ibid., 120, no. 524); [EX] F(iglinis) PATERNI (ibid., 201, no. 1289); F(iglino) SCIM/NIANO (ibid., 248, no. 1579). See in general J.J. Paterson, '"Salvation from the Sea" : amphorae and trade in the Roman West', *JRS*, 72, 1982, 146-157, esp. 155.

53. G. Pucci, art. cit. (n. 23), 111.

54. A. Carandini, 'Pottery and the African economy' in P. Garnsey, K. Hopkins & C.R. Whittaker (ed), op. cit. (n. 16), 145-62, esp. 149, 152, 155-8.

55. I am grateful to Dr. J.C. Faria for kindly showing me examples in the Museu Municipal at Alcácer do Sal and to Dr. F. Reiner and A. Carvalho for doing likewise at the Museu do Mar, Cascais, and for their hospitality.

56. J.J. Paterson, loc. cit. (n. 52).

57. See K. Hopkins, loc. cit. (n. 15).

58. For prices of *garum* see R.I. Curtis, 'In defense of garum' *CJ*, 78, 1983, 232-40.

59. D. Colls, C. Domergue, F. Laubenheimer & B. Liou, 'Les lingots d'étain de l'épave Port-Vendres II', *Gallia*, 33, 1975, 61-94.

60. V. Kockel, 'Archäologische Funde und Forschungen in den Vesuvstädten. II', *AA*, 1986, 443-569, esp. 476-7 for garum shop; cf. R.I. Curtis, 'The garum shop of Pompeii (I.12.8)', *Cron. Pomp.*, 5, 1979, 5-23.

61. For detailed references see Table 6.12.

62. *Dig.* 14.2.2.2, on which see J.A. Crook, *Law and Life of Rome*, London, 1967, 224-225.

63. For Baetican amphorae see M. Maia, 'Marcas em anforas da forma Dr. 20 da Tróia, Setúbal', *AP*, iii, 7-9, 1974-1977, 355-358; for African : M. Maia, 'Contribuição para o estudo das anforas romanas de Tróia - ânforas do tipo "Africano Grande"', *Setúbal Arq.*, 1, 1975, 155-158.

64. A.J. Parker, art.cit. (n. 6), 39.

65. contra, for example, J. M. Blazquez, *Estructura economica y social de Hispania durante la anarquia militar y el Bajo Imperio*, Madrid, 1964.

CHAPTER SEVEN

INDUSTRY AND THE PROVINCIAL ECONOMY

Throughout this study of mining and fish processing it has been difficult to draw any secure conclusions on any quantitative aspect of the subject. The evidence, as is common, if not quite universal, in studies of the economy of the Roman Empire, just will not allow such an approach. Thus it is difficult to assess the scale of production in both the mining and fish processing industries. But what is *a priori* likely is that during the period of the Roman administration of Lusitania there was an increase in the scale of production of both industries, compared to the periods before and after Roman administrative and economic control. It is even possible that fish-processing was only first practised on a wide scale in the first century A.D., and could therefore be seen as an economic response to the creation of the Roman province. However, even though the archaeological evidence is as yet lacking for earlier fish-processing, it would be rash to claim that it simply was not practised before the Roman occupation. Mining certainly did take place in the Bronze and Iron Ages in Lusitania; it was the lure of precious metals that drew Phoenician traders to this westernmost part of Europe.

As for the later history of these two industrial processes, there is scant evidence that mining was practised under the Suevi and the Visigoths who came to ravage and then control this part of the Iberian peninsula from 409 onwards. (1) But the Suevi did mint gold, and much later silver, coins at a relatively high number of Lusitanian mints. Coins minted at Emerita, for instance, enjoyed a wide circulation throughout the peninsula. (2) Gold and silver plate, won as booty or remitted as tax, as well as used Roman coins, were no doubt melted down to produce bullion for this coinage, but it would make sense if the Suevi and Visigoths had continued to operate some of the Roman mines of the area, to ensure that there was always a sufficient supply. References in the writings of Isidore of Seville to alluvial gold and its exploitation might hint at continued production after the Romans, but Isidore was steeped in the Roman literary tradition, and his phraseology at the crucial points bears too many similarities with Pliny the Elder's to give his testimony any independent value for mining in his own period. (3) Penalties of hard labour in the mines still occur in the Visigothic law code, but this may reflect its frequent imitation of Roman legal codes rather than the continued exploitation of mines. (4)

As for fish products, the disappearance even of the later forms of amphorae after the first decades of the fifth century seems to fit well into the context of the Suevic invasion. But there are hints that garum was still exported to Merovingian Gaul in the time of Gregory of Tours. (5) Futhermore, the coming of the Suevi did not extinguish the presence of the Hispano-Roman aristocracy in the west of the peninsula. Are their dietary tastes likely to have undergone radical changes in the wake of political upheaval ? But on the whole the evidence seems to point to the period of the Roman administration of Lusitania as that which saw the greatest extent and scale of production in the mines and in fish processing.

It is not difficult to see why this should have been so. First, the demand was both metals and fish-products was high under the Romans. Metals were needed for coinage, weapons and for the vastly increased building programmes that accompanied the creation of the Roman province. New cities such as Augusta Emerita would have needed enormous quantities of iron and lead for their construction; more precious metals would have been required for such cities' adornment, and for luxury products for the new provincial aristocracy. Existing proto-urban communities such as Conimbriga or Sellium (Tomar) underwent ambitious urbanization programmes immediately after their incorporation into the province of Lusitania. (6) Fish-sauces were a major ingredient of a typically "Roman" dietary pattern. The influx of Roman colonists, and, perhaps more significantly, the Romanization of the native aristocracy would have led to a much higher consumption of the more choice kinds of product. The cheapness and wider availability of the lower grade sauces would have encouraged the lower classes to use them as a salt substitute, and as relish for their cereal and pulse-based diet.

Secondly, various structural elements of the Roman economic system, which would have been introduced into the local economy in the wake of the creation of the province, would have made larger-scale "industrial" enterprises more attractive. In short, there would have been a development in the sophistication of the substructures of the economy, which would have allowed, or even encouraged, economic growth. (7) If we accept that there was economic growth, "industrial" activities are likely to have increased not only in the number of units of production to be found in the province, but also in the size of some of these units; it is possible, as we have seen, that the nucleation of units of production at Tróia was the physical manifestation, visible archaeologically, of "economies of scale" being introduced by those who controlled such "industrial" enterprises. In short, the introduction of Roman administration, with all that that entailed (such as increased urbanization, a changeover to a monetized economy (at least in the towns), the creation of a provincial aristocracy that de rigueur adopted a lifestyle of conspicuous consumption, usually based in the cities), would have led to what may be termed "economic growth", which in turn encouraged "industrial" activities.

It is also possible that a change in the organization of labour, involving the introduction, or development, of the slave mode of production, further encouraged the growth of industrial concerns. Inscriptions certainly attest the presence of slaves and freedmen and freedwomen in various fish-processing establishments in the province, while slaves clearly formed part of the labour pattern in the exploitation of the mines at Vipasca, as the legal texts therefrom attest. That there was a quantitative increase in the use of slaves in Lusitania after the coming of the Romans may be a priori likely (especially in a part of the Roman Empire where there was not only a protracted military struggle for Roman supremacy and hence probably a good supply of slaves, but also a fairly widespread settlement of Roman veterans who would likely have been keen to use slave labour to help cultivate their newly won land), but still needs to be proven. Conclusions based on quasi-statistical analyses of the inscriptions of an area remain largely unconvincing. (8) The full extent, or otherwise, of slave labour in the pre-Roman Celticized period of the later Iron Age needs to be established, before the impact of Roman settlement in the area can be assessed. For other parts of the Empire it has been argued that the Romans were usually keen to leave patterns of labour relations as they had been before their arrival. But if there was an increased use of slaves in agriculture, a landowner who was also involved in industrial concerns, such as mining and fish-processing, would certainly have used these slaves in his industrial enterprises.

Roman rule over Lusitania, as over all other provinces, demanded that there was a significant surplus produced to satisfy the demands of Roman taxation. This demand for surplus must have led to a significant increase in the scale of both mining and fish-processing in the province. Surplus production could then be traded to obtain coinage with which to pay taxes owed to Rome, or rent due to local landowners. (9) The first mechanism for the disposal of this surplus would have been to supply essentially local markets, which would have been the main horizons for most producers in Lusitania. But if a significantly larger surplus had been produced (either in a good year, or if one controlled a relatively large "industrial" concern), then one's thoughts might have turned to export, and overseas markets. The expense of shipping was a major constraining factor on all but the very rich. (10) As Romanization spread through the Empire, the demand for metals and fish preserves increased. As the province became more integrated into the economic pattern of the Empire, it no doubt helped to meet this demand. The finds of Roman amphorae on the route to Rome are indicative of this. In addition the Roman army was a major consumer of both products, especially metals for its arms and armour. There was only one legion permanently stationed in the Iberian peninsula, Legio VII Gemina based at León in the north-west. With the great improvements in the road system, and the increase in, and regularisation of, maritime traffic, Lusitania no doubt helped to keep it supplied. Baetican, and later north African, olive-oil clearly went a long way towards satisfying the demand of the city of Rome and the Roman armies on the Rhine frontier for this product. Since Lusitanian goods were often shipped together with those of Baetica and north Africa, as we have seen, (11) it would be interesting to know to what extent Lusitanian fish-products fulfilled a similar function. The evidence, as it now stands, suggests that Lusitania played only a secondary role in supplying these major Roman markets.

Even if the contribution of Lusitanian goods to the supplying of Rome, and the Roman army, was not as great as that of Baetica or north Africa, the creation of, and increased production in, industrial concerns would have been of great importance to the local, Lusitanian economy and to the Lusitanian upper-classes with the wealth necessary to invest in such concerns. However, an even greater stimulus to increased production will have been caused by the demands of Roman taxation. Hopkins has recently argued that the burden of taxation led to an increase in trade in the Roman Empire, especially in the first two centuries A.D. (12) The close and inter-locking relationship between both mining and garum production and the land is especially relevant here. A Lusitanian landowner would have always been looking to produce a sufficient surplus not only to continue living in a style worthy of, and often demanded by, his social status, but also to use it for the payment of taxes to the Roman government. Agriculture was probably his major concern, but if the necessary resources were available to him, subsidiary "industrial" activities could also have been carried out on his estates, or he could have used some of his wealth derived from agriculture for the exploitation of minerals in State-run mining districts.

To what extent taxes were paid in kind in the early Empire is unclear. Pliny reports that certain poorer Lusitanians paid part of their taxes with the dye extracted from the kermes insects that flourished on a particular kind of oak (Quercus coccifera) common especially around Emerita. (13) Similarly, the surplus extracted from Lusitanian mines or fish-processing establishments could either have been handed over directly to the provincial procurator in payment of taxes in kind or could have been sold on the local market or occasionally to an overseas trader, and the money thus received then handed over to the provincial procurator as tax in form of Roman coinage. Industrial activity was just one aspect of provincial life that was radically altered by the annexation of Lusitania as a province under Augustus; the increase of such activities also unknowingly helped to spread Roman coinage, law and business habits and, more generally, Roman language, customs and culture more widely throughout the province.

NOTES : CHAPTER SEVEN

1. E.A. Thompson, The Goths in Spain, Oxford, 1969, 135, n. 4.

2. For these mints see X. Barral i Altet, La circulation des monnaies sueves et visigotiques : contribution à l'histoire économique du royaume visigot (Beihefte der Francia, 4), Munich, 1976.

3. M.C. Diaz y Diaz, 'Metales y mineria en la epoca visigotica a traves de Isidoro de Sevilla' in La Mineria Hispana e Iberoamericana, León, 1970, 261-274.

4. Lex Visigothorum 2.14.1; 9.10.4; 10.19.4.

5. P.D. King, Law and Society in the Visigothic Kingdom, Cambridge, 1971, 195 (with refs.).

6. Conimbriga : J. de Alarcão & R. Etienne, FC I : L' Architecture, 2 vols., Paris, 1977; Sellium : S. da Ponte, 'Tomar - antiga Sellium', Arqueologia (Porto), 11, jun. 1985, 112-114.

7. K. Hopkins, 'Economic growth and towns in classical antiquity' in P. Abrams & E.A. Wrigley (ed.), Towns in Societies, Cambridge, 1978, 35-77.

8. For example : G. Pereira Menaut, 'El numero de esclavos en las provincias romanas del Mediterraneo occidental', Klio, 63, 1981, 373-99.

9. K. Hopkins, 'Taxes and trade in the Roman Empire', JRS, 70, 1980, 101-25.

10. K. Hopkins, 'Models, ships and staples' in P. Garnsey & C.R. Whittaker (ed.), *Trade and Famine in Classical Antiquity* (= Cambridge Philological Society, Suppl. Vol. 8), Cambridge, 1983, 84-109.

11. See Chapter Six, pp. 180-88.

12. art. cit. (n. 9).

13. Plin. *NH* 9.65.141; 16.12.32

A P P E N D I C E S

APPENDIX I. ROMAN MINES OF LUSITANIA

NOTE ON ABBREVIATIONS

In addition to the abbreviations adopted throughout this work, the following are used in this Appendix :

AMA = S.P.M. Estacio da Veiga, Paleoethnologia : antiguedades monumentães do Algarve (4 vols.), Lisbon, 1886-1891.

CAS = J. Maluquer de Motes, Carta arqueologica de España : Salamanca, Salamanca, 1956.

CME = Catalogo monumental de España

RME = O. Davies, Roman Mines in Europe, Oxford, 1935.

J.C. Allan, Considerations = Considerations on the Antiquity of Mining in the Iberian peninsula, London, 1970.

*C. Domergue (forthcoming) = Catalogue des mines et des fondéries antiques de la péninsule ibérique, Madrid

La mineria = La Mineria hispana e ibero-americana, (7 vols.), León, 1970.

J.M. Maclaren, Gold = Gold : its geological occurence and geographical distribution, London, 1908.

*F. J. Sanchez Palencia (forthcoming) = 'La explotación del oro en la Hispania romana : sus inicios y precedentes' I Congreso Internacional de Mineria antigua en el mundo mediterraneo, Madrid, oct. 1985 (forthcoming)

* I am most grateful to these authors for allowing me to consult these works prior to publication.

Note also the following Portuguese administrative terms :

f. = freguesia; c. = concelho

A. CONVENTUS PACENSIS

DISTRICT OF FARO

P 01 CERRO DE CENAFRECHAL, MORRAÇÃO — Manganese
f. Bordeira; c. Aljezur

Possible Roman mine, with Roman coins.

Ref. : AMA III, 42-43, 78-79; ARA I, 321.

P 02 MARGALHO E PENEDO — Cu
f. Na.Sra. da Alva; c. Aljezur

Shaft and drainage galleries reported.

Ref. : AMA III, 39-41, 78-9; RME, Map IIIc, no. 40.

P 03 VIDIGAL — Cu
c. Aljezur

Possibly worked in Roman period.

Ref. : ARA I, 315.

P 04 ARREGATA — Cu
c. Aljezur

Mine worked from prehistoric to Roman times.

Ref. : ARA I, 316.

P 05 HERDADE DA CORTE DO SOBRO, SERRA DO ROCIO — Fe

f. & c. Aljezur

Open face ancient working. Copper wedges and Roman silver coins (unidentified).

Ref. : AMA III, 41-42, 78-79; ARA I, 316; RME, Map IIIc, no. 39; J. de Alarcão, Portugal romano, 131, 139.

P 06 ADUALHO (4 km W. of Monte Molião) Fe
c. Lagos (+ FOUNDRY)

Four shafts with iron slag at bottom.

Ref. : AMA III, 44-45, 78-79; RME, Map IIIc, no. 42, Rev. Guim. 62, 1952, 134.

P 07 MONCHIQUE Cu
c. Monchique

Ref. : D.F. de Almeida, La mineria, 214.

P 08 CISTERNA DOS CÃES Cu
f. Sé; c. Silves

Shaft of 41m; stone mallets, iron mallet, many fragments of Roman tegulae.

Ref. : AMA III, 45-48, 78-79; ARA II, 99-100, RME, Map IIIc, no. 49.

P 09 SÃO ESTEVÃO (5 km NE of Silves) Cu
f. Sé; c. Silves

Bronze Age mallets, Roman iron picks, lucerna (C4 A.D.), coins (C4).

Ref. : AMA III, 48-57, 78-79; ARA II, 111-114; RME, Map IIIc, no. 48; J. Alarcão, Portugal romano, 131, 134; C. Domergue, POR 27.

P 10 PICO ALTO (3 km E. of S. Bartolomeu de Messines) Cu Fe
f. Sé; c. Silves

Ref. : AMA III, 57-58, 78-79; RME, Map IIIc, no. 50.

P 11 CERRO DO MONTE ROSO Cu
f. S. Bartolomeu de Messines; c. Silves

Traces of ancient mines and mining tools.

Ref. : AMA II, 364; ARA II, 117; M.E.H. Henriques Costa, Actas do I Congreso Nacional de Arqueologia Lisboa 1958, II, 77.

P 12 VALE DE PEGAS Cu
f. Paderne; c. Albufeira

Two ancient shafts (10m deep). Bronze Age axes.

Ref. : AMA III, 58-59, 78-79; AMA IV, 89; RME, Map IIIc, no. 1; L. Monteagudo, Caesaraugusta, 4, 1954, 93.

P 13 ATALAIA DE ALTE Cu
f. Alte; c. Loulé (+ FOUNDRY)

Roman working and copper ingot (now in Museu Nacional de Arqueologia e Etnologia, inv.no. 10219).

Ref. : AMA III, 59-61, 78-79; ARA II, 143; RME, Map IIIc, no. 53; D.F. de Almeida, La mineria, 214.

P 14 ALAGOAS Fe
c. Loulé (+ FOUNDRY)

Traces of Roman workings, iron slag and Roman ceramics (unspecified).

Ref. : ARA II, 159.

P 15 VENDINHA DO ESTEVAL Cu
f. Querença; c. Loulé

Roman drainage gallery, Bronze Age tools.

Ref. : AMA III, 61-64, 78-79; RME, Map IIIc, no. 54.

P 16 OSSONOBA (modern FARO) Sn

Arab writer (Maccari) refers to tin mines.

Ref. : L. Monteagudo, Caesaraugusta, 4, 1954, 91; A. Tovar, Iberische Landeskunde, II.2, 207.

P 17 PEDRA DO LEÃO Cu
f. Sta. Catharina; c. Tavira

Ref. : AMA III, 65, 78-79; RME, Map IIIc, no. 58.

P 18 MALHADA DO NOBRE Cu
f. Sta. Catharina; c. Tavira (+ FOUNDRY)

Roman finds common in the area of the mine, including three pieces of slag.

Ref. : AMA III, 65, 78-79; ARA II, 294; RME, Map IIIc, no. 57.

P 19 SERRA DA MINA E CONCEIÇÃO, HERDADE DA MEDRONHEIRA — Cu
f. Odeleite; c. Castro Marim

Roman shaft (15m deep) and drainage gallery.

Ref. : AMA III, 65-67, 78-79; D.F. de Almeida, *La mineria*, 214.

P 20 PEGO DE SÃO DOMINGOS — Fe ?
c. Castro Marim

Unconfirmed reports of Roman mine.

Ref. : ARA II, 365.

P 21 SERRO DAS PEDRAS E DA GALLINHA — Cu (+ FOUNDRY)
f. Vaqueiros; c. Alcoutim

Many ancient working and slag reported.

Ref. : AMA III, 74-75, 78-79; RME, Map IIIc, no. 63.

P 22 HERDADE DA MALHADA, FORRA MERENDAS — Cu (+ FOUNDRY)
f. Vaqueiros; c. Alcoutim

Much open-face Roman working & slag from Serro das Borras in vicinity, 15 km from mouth of Ribeira de Odeleite.

Ref. : AMA III, 73-74, 78-79; ARA II, 391; RME, Map IIIc, no. 61.

P 23 COVA DOS MOUROS Cu
c. Alcoutim

Ancient shaft (30m deep) reported. Roman Republican coin hoard found nearby at Monte de Sodes.

Ref. : AMA III, 69-73, 78-79; ARA II, 383, 389; RME, Map IIIc, no. 62; J. Alarcão, Portugal romano, 131, 134.

P 24 LABORATO Cu
f. Martim Longo; c. Alcoutim

Roman mines reported in same Cerro as P25.

Ref. : AMA III, 69, 78-79; ARA II, 383.

P 25 DAROEIRA (or AROEIRA) Cu
f. Martim Longo; c. Alcoutim

Much evidence of Roman working. Denarius of Antoninus Pius suggests that mine still operative in C2.

Ref. : AMA III, 68-69, 78-79; ARA II, 383, 390; RME, Map IIIc, no. 65; J. Alarcão, Portugal romano, 131, 134.

P 26 SERRA DA MINA Cu

f. Martim Longo; c. Alcoutim

Roman mines reported c. 1500 m NE of Martim Longo.

Ref. : RME, Map IIIc, no. 60; C. Domergue, POR 23.

P 27 SERRO DOS RELIQUIAS Cu

f. Giões; c. Alcoutim

Roman mines reported on right bank of the ribeira do Vascão.

Ref. : ARA II, 383, 391.

P 28 LUTAO Cu

c. Alcoutim

Ref. : ARA II, 383.

P 29 CORTES PEREIRA Cu Ant

f. & c. Alcoutim

No direct evidence of ancient working, but plenty of traces of Roman occupation of this mining area 5 km from Alcoutim, 1 km SW of river Guadiana.

Ref. : AMA III, 67-68, 78-79; ARA II, 383, 386; RME, Map IIIc, no. 64.

DISTRICT OF BEJA

P 30 BRANCANES Cu
c. Almodovar (+ FOUNDRY)

Three ancient shafts (20m deep). Coins of Augustus, Vespasian and Probus, suggesting the mine was in operation at least from start C1 A.D. - 2nd half of C3.

Ref. : C. Domergue (forthcoming) POR 3

P 31 HERDADE DO MONTINHO Ag
c. Ourique (+ FOUNDRY)

Shafts (15m) discovered on right bank of the ribeira das Ferrarias, and mining instruments.

Ref. : C. Domergue (forthcoming) POR 11

P 32 NEVES CORVO, SOMINCOR Cu
c. Castro Verde Ag
Pb

Mines reported in vicinity of Iron Age agricultural settlement (with Punic amphorae, Greek pottery and imported glassware), and a fort of the Augustan period, and a later Roman villa.

Ref. : M. Maia, Inform. arq., 5, 1982-3, 40-41.

P 33 METALLUM VIPASCENSE (Algares & S. João do Deserto)
c. Aljustrel Cu/Ag/Au?/Fe?/quarries
(+ FOUNDRIES)

Important copper and silver (and possibly gold and iron mines and stone quarries) worked in Bronze Age and then from 2nd half of C2 B.C., with slag surviving in large quantity (c. 3,000,000 tonnes) from latter part of C1 B.C. onwards. Mines still being worked in the later C3 & C4 A.D., after some disruption in late C2/early C3. Maximum depth of shafts some 30m. Many shallower workings (often sunk in pairs). Various items of mining machinery (windlasses and frames) and equipment (miners' sandals, caps, baskets (all of esparto grass), picks, hammers). Two ingots - one copper, one lead - discovered. Mining settlment and cemetery also excavated at Valdoca. Also two legal texts regarding the administration of the mines found in slag heaps.

Refs. : C. Domergue, La mine antique d'Aljustrel, 1983;
RME Map IIIc, no. 36;
A. Viana, O. da V. Ferreira, R.F. d'Andrade, Arq. de Beja, 13, 1956, 3-20;
R.F. d'Andrade, Actas do I Congreso Nacional de Arqueologia Lisboa 1958, II, 273-285;
J. & A. de Alarcão, Conimbriga, 5, 1966, 7-104 (Valdoca cemetery);
C. Domergue & R.F. d'Andrade, Conimbriga, 10, 1971, 99-116;
R. Parreira, Inform. arq., 4, 1981, 135-44; Arq. de Beja, iii, 1, 1984, 83-106 (mining settlement).

P 34 JULIANA Cu
f. Santa Vitoria (+ FOUNDRY)

Ancient shafts and galleries; stone hammers, copper and bronze axes, attesting pre-Roman working; uncertain if exploited under the Romans.

Ref. : AMA, III, 18 & n.2; RME, Map IIIc, no. 69; C. Domergue (forthcoming), POR 6.

P 35 SERRO DE COVOES, ZAMBUJAL — Fe
c. Alvito

Large number of shafts reported together with an iron pick.

Ref. : D.F. de Almeida, La mineria, 217 & fig. 1; C. Domergue (forthcoming), POR 4.

DISTRICT OF SETÚBAL

P 36 SERRA DA CAVEIRA — Cu, Ag, Au
c. Grândola (+ FOUNDRY)

Enormous number of shafts, three galleries, a drainage channel and much slag (c. 300,000 tonnes). Copper axe suggest Bronze Age exploitation. Roman *lucernae* suggest working from the Augustan period into C4 A.D. Thin-walled ware suggests C1 A.D. working.

Ref. : J.L. de Vasconcelos, *AP*, 19, 1914, 310-311; *Religiões da Lusitania*, III, 484-486; *De terra em terra*, II, 95-98; *RME*, Map IIIc, no. 68; D.F. de Almeida, *La mineria*, 213; J.C. Allan, *Considerations*, 7; C. Domergue (forthcoming), POR 36.

P 37 ADIÇA (near Costa da Caparica) — Au

Alluvial gold referred to by Arab writer, Edrisi. Probably worked by the Romans.

Ref. : J.C. Allan, *Considerations*, 2; D.F. de Almeida, *La mineria*, 213; J.M. Maclaren *Gold*, 1908, 143.

P 38 ALMADA Au

Alluvial gold worked by the Romans.

Ref. : RME, Map IIIc, no. 13 & p. 106.

DISTRICT OF EVORA

P 39 SERRA DOS MONGES Cu
c. Montemor-o-Novo Fe

Roman gallery. Neolithic stone axes, Roman *lucernae* of C1 A.D. and unspecified amphorae.

Ref. : C. Domergue (forthcoming) POR 22a; J. Alarcao, *Portugal romano*, 131, 139.

P 40 NOGUEIRINHO Cu
c. Montemor-o-Novo Fe

Ref. : C. Domergue (forthcoming) POR 22.

P 41 MINA DO BUGALHO (or BAGALHO) Cu
c. Alandroal

Three seams of metal exploited in Roman times. Various mallets found on the site.

Ref. : RME, Map IIIc, no. 70; C. Domergue (forthcoming) POR 21.

P 42 VILA VIÇOSA Cu

Ref. : RME, 269 & Map IIIb, no. 25.

DISTRICT OF PORTALEGRE

P 43 PAÇO DO CONCELHO — Cu
c. Elvas

Ref. : AP, iii, 1, 1967, 120.

P 44 HERDADE DA TINOÇA — Cu
c. Arronches

Probable Roman working.

Ref. : C. Domergue (forthcoming) POR 29.

P 45 COVA DA MOURA — Au, Ag, Pb
c. Marvão

Gallery of ancient mine reported.

Ref. : J. Conceição Neves, Conimbriga, 11, 1972, 5-33, esp. 5.

P 46 SEVER VALLEY — Au

Roman alluvial workings.

Ref. : D.F. de Almeida, La mineria, 217.

P 47 VALE DOS MOUROS, CONHAL Au
c. Nisa

8 drainage channels reported on right bank of the ribeiras da Zorna and de Nias, as well as alluvial platform.

Ref. : Inform. arq., 5, 1982-83, 19-20.

P 48 PORTAS DO RODÃO Au
c. Nisa

Traces of alluvial workings on Tagus between Serrinha and Serra de S. Miguel.

Ref. : Inform. arq., 5, 1982-83, 19.

ROMAN SLAG SITES IN THE CONVENTUS PACENSIS

SL01	Praia do Amado c. Aljezur	AMA III 78-9; RME Map IIIc, no. 41
SL02	Torrejão c. Alzejur	AMA III 78-9; RME Map IIIc, no. 38
SL03	Serro das Alforrobeiras c. Vila do Bispo	AMA III 78-9; RME Map IIIc, no. 44; ARA II 42; AP, 23, 1917, 130
SL04	Paul c. Lagos	AMA III 78-9; RME Map IIIc, no. 43
SL05	Serro do Pezo c. Lagos	AMA III 78-8; RME Map IIIc, no. 45

SL06	Serro do Haver c. Lagos	AMA III 78-9;
SL07	Cruzinha c. Lagos	AMA III 78-9; ARA II, 7; RME Map IIIc, no. 46
SL08	Palmeirinha c. Lagos	AMA III 78-9; RME Map IIIc, no. 47
SL09	Itheu do Rosario c. Silves	AMA III 78-9; ARA II, 90; RME Map IIIc, no. 51
SL10	Zambujal c. Silves	AMA III 78-9
SL11	Terras Velhas c. Silves	Actas do I CNA Lisboa, 1958, Lisbon, 1970, II, 82.
SL12	Serros Altos c. Albufeira	AMA III 78-8; ARA II, 400, RME Map IIIc, no. 52
SL13	Milreu c. Faro	AMA III 78-9; RME Map IIIc, no. 55
SL14	Campo da Trindade c. Faro	AMA III 78-9.
SL15	Amendoal c. Faro	AMA III 78-9; RME Map IIIc, no. 56
SL16	Poço do Valle c. Tavira	AMA III 78-9.
SL17	Torre de Ares c. Tavira	AMA III 78-9; RME Map IIIc, no. 59
SL18	Sitio dos Olhos c. Castro Marim	AMA III 78-9

B. CONVENTUS EMERITENSIS

PROVINCE OF BADAJOZ

E 01 LA PERDIZ Ag/Pb/Cu
Oliva de Mérida (+ FOUNDRY)

Ref. : C. Domergue (forthcoming), BA 56.

E 02 ARROYO DE LAS MINETAS Ag
Oliva de Mérida Pb

Ref. : C. Domergue (forthcoming), BA 57.

E 03 ARROYO CHAMAZO Ag
Oliva de Mérida

Ref. : C. Domergue (forthcoming), BA 55.

E 04 EL JUDIO Ag
5 km ESE of Mérida

Ref. : C. Domergue (forthcoming), BA 54.

E 05 CORTIJO DEL SANTO, Montemolin — Ag

Several Roman mines reported from the Sierra de Estrella in enclave of territory of Emerita, the capital of Lusitania.

Ref. : P. Sillières, MCV, 18, 1982, 446 & n. 37.

E 06 TRES ARROYOS 8 km NW of Albuquerque (+ FOUNDRY) — Sn

Tin mines worked under Romans. Terra sigillata hispanica dated to end C1/start C2 A.D. Possible source for ingots found on Port-Vendres II wreck.

Ref. : C. Domergue (forthcoming), BA 1; Gallia, 1975, 85; Archaeonautica, 1977, 16 (ingots).

E 07 LEONCILLO Villa del Rey — Sn

Slag reported from vicinity of Roman villa.

Ref. : J.-G. Gorges, Les villas hispano-romaines, 197, sv. BA 32.

E 08 LA CODOSERA — Au

Report of mines (open-cast and shafts), traces of slag, granite millstones for the treatment of ore near Portuguese frontier, 15 km west of Albuquerque, 20 km south of S. Vicente de Alcántara.

Ref. : F.J. Sanchez Palencia (forthcoming), no. 22.

PROVINCE OF CÁCERES

E 09 S. VICENTE DE ALCÁNTARA Au

Possible Roman gold mines - open-cast.

Ref. : F.J. Sanchez Palencia (forthcoming), no. 23.

E 10 VALENCIA DE ALCÁNTARA Au

Possible Roman gold mines - open-cast.

Ref. : F.J. Sanchez Palencia (forthcoming), no. 23.

E 11 RIVER SEVER Au

Possible alluvial placers.

Ref. : F.J. Sanchez Palencia (forthcoming), no. 23.

E 12 MEMBRIO Au

Report of possible Roman gold mines - open-cast.

Ref. : F.J. Sanchez Palencia (forthcoming), no. 23.

E 13 RIVER SALOR Au

Possible Roman alluvial placers.

Ref. : F.J. Sanchez Palencia (forthcoming), no. 23.

E 14 PERALES DEL PUERTO Sn
N. of Coria

Tin mines worked in Roman period.

Ref. : M. Almagro Gorbea, <u>El Bronce final</u>, Madrid, 1977, 8.

E 15 SIERRA DEL MORO, CADALSO & HERNAN PEREZ Au

Possible Roman gold mines with galleries reported.

Ref. : J. Perez Regodon, 'Investigaciones auriferas en la provincia de Cáceres', <u>BIGME</u>, 80, 1969,, 131-45, esp. 142; F.J. Sanchez Palencia (forthcoming), no. 25.

E 16 MINA CHIVOTE, CALZADILLA Au

Open-cast workings on the right bank of the arroyo de Patano, a tributary of the river Arrago, stretching for some 78m. 1 shaft also visible. Report (imprecise) of some Roman ceramics, tools and an inscription.

Ref. : J. Perez Regodon, art. cit., esp. 133 & 136; F.J. Sanchez Palencia (forthcoming), no. 24.

E 17 ALGIBE Ag
15 km S of La Aliseda (+ FOUNDRY) Fe

Roman silver/iron mines. Bronze and silver coins of Trajan found.

Ref. : RME Map IIIb, no. 17; C. Domergue (forthcoming), CC 1.

E 18 VALDEFLORES Sn
5 km from Cáceres

Ref. : L. Monteagudo, Caesaraugusta, 4, 1954, 90.

E 19 PLASENZUELA Ag
(+ FOUNDRY) Cu
Pb

Important mining district stretching for 6 km, NW of modern village. 2,500 to 3,000 Roman workings visible at end of C19. Granite washing table in situ. Finds include mining picks, hammers, Roman tiles, amphorae, lamps (one dated to 1st half C1 B.C.). Inscr. (C1 A.D.) : L. Iulius Ibarr...Lascivi f. Iron tools & lead tessera now preserved in Museo Arqueologico Nacional, Madrid.

Ref. : CME, Cáceres, I, 192; C. Domergue (forthcoming), CC 2.

E 20 BERZOCANA Cu

Deep mining shaft reported 2 km E of Rio Garciaz with circular houses nearby.

Ref. : XIII Congreso Nacional de Arqueologia, Huelva 1973, Zaragoza, 1975, 472-473.

PROVINCE OF TOLEDO

E 21 SANTA QUITERIA Ag
Sevilleja de la Jara Pb

Iron hammers and picks and Roman lucernae reported from silver/lead mines.

Ref. : C. Domergue (forthcoming), TO 1.

E 22 NAVA DE RICOMALILLO Au
(+ FOUNDRY)

On the northern fringe of the Montes de Toledo, 43 km from Talavera de la Reina. Three strains of gold worked in Roman period (La Rica, Ingeniera, Eugenia). Fragments of Roman tiles, ceramics, washing tables, mill-stones and slag. Also four Latin inscriptions.

Ref. : BRAH, 30, 1897, 426-428; BRAH, 36, 1900, 175-6, 445-6; BRAH 43, 1903, 536; C. Domergue (forthcoming), TO 2.

PROVINCE OF SALAMANCA

E 23 SERRADILLA Sn

Roman mine with galleries and shafts reported. Finds of Roman coarseware and mining lamp.

Ref. : T. de la Fuente, AEA, 28, 1955, 317-318.

E 24 LOS CAVENES, EL CABAÇO Au

Six or possibly seven Roman reservoirs and water channels found in vicinity of El Cabaço indicate a substantial exploitation of alluvial gold. The mining technique seems very similar to that used by the Romans in the north-west of Spain.

Ref. : CME, León, 97; JRS, 62, 1972, 74 & n. 48; C. Domergue (forthcoming), SA 2; F.J. Sanchez Palencia (forthcoming), no. 30.

E 25 GUIJELO Ag Pb

Silver/lead mines reported from 1500m SSE of Guijelo.

Ref. : C. Domergue (forthcoming), SA 4.

E 26 ESTRELLA Pb
Aldeavieja de Tormes

Lead mines worked in Roman period 800m N of Aldeavieja de Tormes.

Ref. : C. Domergue (forthcoming), SA 1.

E 27 SAN PEDRO DE ROSADOS Sn

Tin mines worked in Roman period.

Ref. : L. Monteagudo, Caesaraugusta, 4, 1954, 90.

E 28 BERNOY Sn

Roman tin mines immediately to the south-west of E 26.

Ref. : L. Monteagudo, Caesaraugusta, 4, 1954, 90.

E 29 GOLPEJAS Sn
23 km W of Salamanca

Tin mines worked in Bronze Age and possibly in Roman times.

Ref. : C. Domergue (forthcoming), SA 3.

E 30 TERRUBIAS Sn

Ref. : L. Monteagudo, loc.cit.

E 31 PERENA Sn

Ref. : L. Monteagudo, loc.cit.

E 32 VILLARES DE YELTES Sn

Ref. : L. Monteagudo, Caesaraugusta, 4, 1954, 90.

E 33 LAS MERCHANAS Sn
Lumbrales

Military headquarters in C3 A.D. to control tin mining of region.

Ref. : J. M. de Motes, _Pyrenae_, 4, 1968, 119; _CAS_, 86-7; J.M. Blazquez, _La mineria_, 148.

E 34 FREGENEDA Sn

Ref. : L. Monteagudo, _Caesaraugusta_, 4, 1954, 90.

C. CONVENTUS SCALLABITANUS

DISTRICT OF LEIRIA

S 01 PEDROGÃO, POVOA DE COS Fe

Foundry with iron slag attached to Roman villa.

Ref. : J.-G. Gorges, Les villas hispano-romaines, 461, PC 08.

S 02 ARNEIRO, ARNAL Fe

Iron mine close to Roman villa.

Ref. : J.-G. Gorges, Villas, 460 & PC 06.

DISTRICT OF SANTARÉM

S 03 TAGUS VALLEY I : SANTARÉM Au

Roman alluvial workings.

Ref. : J.M. Maclaren, Gold, 143; RME Map IIb, no. 9.

S 04 TAGUS VALLEY II : ALMEIRIM Au

Roman alluvial workings north of Almeirim.

Ref. : J.M. Maclaren, Gold, 143; RME Map IIb, no.

S 05 TAGUS VALLEY III : CONSTANCIA Au
(Confluence of Tagus and Zêzere)

Three Roman terraces stretching for 3 km on R. bank of Tagus; one terrace on L. bank of Zêzere.

Ref. : D.F. de Almeida, La mineria, 212; C. Domergue (forthcoming), POR 34; F.J. Sanchez Palencia (forthcoming), no. 9.

S 06 TAGUS VALLEY IV : ORTIGA, ABRANTES Au
c. Abrantes

Roman terraces on R. bank of Tagus on both sides of the confluence with the Rio das Hortas.

Ref. : C. Domergue (forthcoming), POR 32; F.J. Sanchez Palencia (forthcoming), no. 10.

S 07 TAGUS VALLEY V : MOURISCAS Au
c. Abrantes

Roman terraces on both banks of Tagus.

Ref. : C. Domergue (forthcoming), POR 33.

S 08 TAGUS VALLEY VI : ALVEGA Au

Roman terraces on both banks of Tagus.

Ref. : J.M. Maclaren, Gold, 143; RME Map IIIb, no. 11; F.J. Sanchez Palencia (forthcoming), no. 10.

S 09 MILREU — Au
c. Vila de Rei

Alluvial exploitations (i) 1 km S & SE of Milreu on R. bank of Ribeira de Codogoso, (ii) 3.5 km SE of Milreu at the confluence of the Ribeiras da Galega and de Cades. Also an aqueduct (5 km) for bringing water to wash the gold.

Ref. : J.C. Allan, Boletim de Minas, 2, 1965, 23; J. Alarcão, Portugal romano, 131; C. Domergue (forthcoming), POR 35; F.J. Sanchez Palencia (forthcoming), no. 7.

S 10 RIBEIRA DE PRAÇANA (tributary of Tagus) — Au

Roman alluvial workings.

Ref. : J. Alarcão, Portugal, 133; F.J. Sanchez Palencia (forthcoming), no. 12.

S 11 RIBEIRA OCRESA (tributary of Tagus) — Au

Roman alluvial workings.

Ref. : J. Alarcão, Portugal, 133; F.J. Sanchez Palencia (forthcoming), no. 12.

DISTRICT OF CASTELO BRANCO

S 12 RIO TRAPEIRO Au

Roman alluvial workings on this tributary of the Ocresa near Chão da Va and Salgueiro do Campo.

Ref. : F.J. Sanchez Palencia (forthcoming), no. 13.

S 13 PORTAS DA RODÃO Au
c. Vila Velha da Rodão

Clear traces of Roman exploitation on rivers on two sides of Vila Velha da Rodão.

Ref. : F.J. Sanchez Palencia (forthcoming), no. 14.

S 14 CHARNECA Au
c. Vila Velha da Rodão

Alluvial terraces exploited by Romans discovered near Perais and Serranos, immediately to the east of Vila Velha.

Ref. : F.J. Sanchez Palencia (forthcoming), no. 15.

S 15 CABEÇO DE MOURO Au

Gold workings on the Tagus, 10 km S. of Rosmaninhal.

Ref. : D.F. de Almeida, La mineria, 212.

S 16 ROSMANINHAL Au

Shafts and gallery discovered from Roman gold workings.

Ref. : J.M. Maclaren, Gold, 143; RME Map IIIb, no. 12; D.F. de Almeida, La mineria, 212; J. Alarcão, Portugal, 131, 133; F. J. Sanchez Palencia (forthcoming), no. 19.

S 17 RIO ELGES (or ERGES) Au

Alluvial workings on the Hispano-Portuguese border.

Ref. : J.M. Maclaren, Gold, 143; D.F. de Almeida, La mineria, 212; J. Alarcão, Portugal, 133.

S 18 RIBEIRA DE STA. MARINHA Au
(tributary of Elges)

Roman alluvial workings discovered NE of Segura.

Ref. : D.F. de Almeida, La mineria, 212.

S 19 MONFORTINHO Au
c. Idanha-a-Nova

Roman terraces for alluvial workings on R. bank of Rio Elges at three points near Monfortinho, stretching over an area of 5 km.

Ref. : J.M. Maclaren, Gold, 143; D.F.de Almeida, La mineria, 212; C. Domergue (forthcoming), POR

17; F.J. Sanchez Palencia (forthcoming), no. 20.

S 20 RIO ARAVIL (tributary of Tagus) Au
c. Idanha-a-Nova

Roman alluvial workings.

Ref. : D.F. de Almeida, La mineria, 212; J. Alarcão, Portugal, 133; F.J. Sanchez Palencia (forthcoming), no. 18.

S 21 RIO PONSUL : MONTE POMBAL Au
c. Castelo Branco

Roman alluvial workings stretching 4 km along this tributary of the Tagus either side of the confluence with the Rio Ribeirinho.

Ref. : D.F. de Almeida, La mineria, 212; J. Alarcão, Portugal, 133; C. Domergue (forthcoming), POR 16; F.J. Sanchez Palencia (forthcoming), no. 16.

S 22 RIO PONSUL : IDANHA-A-NOVA Au
c. Idanha-a-Nova

Roman alluvial workings reported.

Ref. : F.J. Sanchez Palencia (forthcoming), no. 17.

S 23 MEDELIM Sn
c. Idanha-a-Nova Pb

Tin/lead mines 15 km S. of Penamacor, possibly worked since Roman times.

Ref. : P. Descamps, _Portugal, la vie sociale actuelle_, 285.

S 24 RIO BASEGUEDA (tributary of Elges) Au
c. Penamacor

Roman alluvial workings.

Ref. : D.F. de Almeida, _La mineria_, 212; J. Alarcao, _Portugal_, 133; F. J. Sanchez Palencia, (forthcoming), no. 21.

S 25 MINA DO PINHEIRO Pb
c. Penamacor Au

Lead mines 4 km SE of Penamacor possibly worked by the Romans.

Ref. : P. Descamps, _Le Portugal_, 285.

S 26 MEIMOA Au
c. Penamacor

Extensive Roman alluvial workings, stretching 5 km along the R. bank of the Rib. de Valdedra. Concentrated workings at Covão do Urso and La Presa. At La Presa terraces 2.5 km long by 1 km wide & Roman coins (36 Republ. _denarii_, 4 of M. Antonius, 17 of Octavian). Estimated that c. 2,500,000 cubic metres of alluvium exploited by the Romans.

Ref. : D.F. de Almeida, _La mineria_, 212; J. Alarcão, _Portugal_, 131, 133-4; M. Lourdes Albertos & M. Pires Bento, _XIV Congreso Nacional de Arqueologia Vitoria 1975_, 1201-1203; C. Domergue (forthcoming), POR 18.

S 27 RIO ZÊZERE : BELMONTE Sn
c. Belmonte

Extensive tin alluvial workings in the upper Zêzere valley. Roman coins found, but dispersed without careful recording.

Ref. : J.C. Allan, <u>Considerations</u>, 28-29; J. Alarcão, <u>Portugal</u>, 139; C. Domergue (forthcoming), POR 15.

S 28 MINA DA GAIA Sn
c. Belmonte

Roman alluvial tin workings.

Ref. : D.F. de Almeida, <u>La mineria</u>, 214; P. Descamps, <u>Le Portugal</u>, 285.

S 29 RIO MAÇAINHAS (tributary of the Zezere) Sn
c. Belmonte

Roman alluvial tin workings. Stone and iron mining tools. Roman coins (1st half C1 A.D.).

Ref. : J.C.Allan, <u>Considerations</u>, 28-29; J. Alarcão, <u>Portugal</u>, 139; C. Domergue (forthcoming), POR 15.

S 30 FERRO Fe
c. Covilhã

Lots of Roman iron slag.

Ref. : M.A.G.C. Rodrigues, <u>Estudos de Castelo Branco</u>, 50, 1974, 67-79, esp. 67.

DISTRICT OF GUARDA

S 31 CABEÇO DAS FRAGUAS Sn/Cu/Fe
c. Guarda

Roman tin mines near Romanized Iron Age hill-forts. Slag from nearby Vale Grande, Quinta de S. Domingos.

Ref. : A. Vasco Rodrigues, 'O castro do Cabeço das Fraguas e a romanização das suas imediações', Beira Alta, 1959 (offprint, no page numbers).

S 32 SABUGAL Cu
c. Guarda Pb

Roman gallery (12m deep); Roman mining axe. Lead weights from here now in Museu Machado de Castro, Coimbra.

Ref. : AMA IV 225; L. Monteagudo, Caesaraugusta, 4, 1954, 72; C. Domergue, POR 28.

S 33 TELHÕES Sn
c. Pero Martins

Roman tin mines 500m from possible villa site. Other tin mines of area also possibly worked in Roman period.

Ref. : M. Maia, AP, iii, 7-9, 1974-1977, 209-213, esp. 211.

S 34 MEDA Pb

Location of Roman Meidubriga. Pliny talks of the Medubrigenses 'qui plumbarii sunt'.

Ref. : Plin. NH 4.22.118 & J. Alarcão, Portugal, 139.

DISTRICT OF VISEU

S 35 SÃO ANTONIO Au
c. Penedono

Gold mines worked in Roman times.

Ref. : D.F. de Almeida, La mineria, 211; J. Alarcão, Portugal, 131, 134; F.J. Sanchez Palencia (forthcoming), no. 29.

S 36 SERRA DA QUERIGA Sn
c. Satão

Tin mine worked in the Roman period. Mining axe.

Ref. : J. Alarcão, Portugal, 121; C. Domergue (forthcoming), POR 46.

DISTRICT OF AVEIRO

S 37 SÃO PEDRO DO SUL Sn
c. S. Pedro do Sul

Tin mines possibly worked under the Romans.

Ref. : H.H. Howarth, Trans. of Ethnol. Soc. of London, n.s., 6, 1867, 81.

S 38 MINA DO BRAÇAL, MALHADA Pb
c. Malhada

Two strains of lead worked in Roman times, between the Rio Mau and the Coval do Mo. Gallery discovered on the R. bank of the Coval do Mo. Two lamps (1st half C1 A.D.), one of ? C2 A.D.

Ref. : D.F. de Almeida, La mineria, 214; J. Alarcão, Portugal, 131, 139; C. Domergue (forthcoming), POR 1.

S 39 ALBERGARIA VELHA Sn
c. Albergaria Velha

Tin mines possibly exploited by Romans.

Ref. : RME Map IIIb, no. 14.

DISTRICT OF COIMBRA

S 40 ALGARES/ALGARVES Au

Alluvial gold possibly obtained from the Rib. do Porto and the Rib. de Algares in Roman times.

Ref. : D.F. de Almeida, La mineria, 212.

S 41 ESCADIA GRANDE : SERRA DA LOUSÃ Au
f. Sibares; c. Gois Ag

Three Roman mines in this area, with galleries and "desmontes" discovered. Roman lamps of 1st half C1 A.D., C1 A.D., C2 A.D. and early Christian lamps of C5 (uncertain if latter from mines). Inscr. C1/2 A.D. to indigenous deity.

Ref. : C. Teixeira, Trab. de Antr. e Etnol., 10, 1945-1946, 243-247; D.F. de Almeida, La mineria, 212; J. Alarcão, Portugal, 131, 133; C. Domergue (forthcoming), POR 20; O. da V. Ferreira, Rev. Guim. 1952, 192-195.

S 42 FURADAS DE POMBEIRO Au
c. Arganil

? Roman galleries discovered 1500m N. of Pombeiro on Rio Alva.

Ref. : J. Alarcão, Portugal, 131, 134 (sceptical); cf. C. Domergue (forthcoming, POR 19, arguing for their exploitation under the Romans).

S 43 COJA Au

Alluvial workings on river Alva in the Arganil valley.

Ref. : F.J. Sanchez Palencia (forthcoming), no. 27.

S 44 SARZEDO Au

Alluvial workings on river Alva in the Arganil valley.

Ref. : F.J. Sanchez Palencia (forthcoming), no. 27.

S 45 RIO CEIRA
c. Gois Au

Roman alluvial workings reported along Ceira at Serpins, Vila Nova de Ceira and Gois.

Ref. : F.J. Sanchez Palencia (forthcoming), no. 28.

APPENDIX II. THE LEGAL TEXTS FROM VIPASCA

VIPASCA A = LEX METALLI VIPASCENSIS

(= *CIL* II 5181 = *FIRA* I, 105)

1. OF THE ONE-PERCENT TAX ON AUCTION TRANSACTIONS

i. The lessee (*conductor*) of the tax on sales held at auction within the district of the Vipasca mines shall receive one percent from the seller except on sales which the procurator of the mines holds on the order of the emperor.

ii. The lessee shall exact one percent of the price for working mineshafts (*putei*) which the procurator of the mines sells.

iii. If, after the auction has begun, everything is knocked down in one lot, the seller shall, nevertheless, be required to pay one percent to the lessee, or to his partner or agent (*conductor sociusve actorve*).

iv. The lessee, or his partner or agent, if he wishes, shall be allowed to stipulate or obtain security.

v. The lessee, or his partner or agent, shall exact one percent also of any amount that is received during an auction.

vi. If anyone has property for sale through the auctioneer and he does not sell it at auction, but sells it himself by contract within ten days of placing it with the auctioneer, he shall, nevertheless, be required to give one percent to the lessee, or his partner or agent.

vii. If anything owed to the lessee, or his partner or agent, in accordance with this section of the regulations, is not given, paid, or secured within the next three days after the debt is incurred, he shall have to pay double.

2. OF THE AUCTIONEERING FEE

i. Whoever leases the auctioneering concession (praeconium) shall provide an auctioneer within the mining district.

ii. On a sale of 100 denarii or less he shall collect a fee of two percent from the seller, on a sale of over 100 denarii, one percent.

iii. Anyone who sells slaves through the auctioneer shall be required to pay the lessee, or his partner or agent [six ?] denarii per head if he sells five or less, three denarii if he sells more than five.

iv. If the procurator of the mines sells or leases any property in the name of the fisc, the lessee, or his partner or agent, shall be required to furnish an auctioneer for the sale of such property.

v. Anyone who posts a list with a view to selling any property whatever shall be required to pay one denarius to the lessee, or to his partner or agent.

vi. The purchaser of mineshafts sold by the procurator of the mines shall be required to pay one percent.

vii. If he does not pay it within three days, he shall have to pay double.

viii. The lessee, or his partner or agent, shall have the right to obtain security.

ix. Anyone who sells mules, asses or horses, male or female, through the auctioneer shall be required to pay three denarii per head.

x. Anyone who places slaves or any other property for sale through the auctioneer and sells it himself by contract within thirty days shall be required to pay the same (or "a fee" ?) to the lessee, or to his partner or agent.

3. OF THE MAINTENANCE OF THE BATHS

i. The lessee of the baths or his partner shall, in accordance with the terms of the lease valid until the forthcoming 30 June, be required to heat the baths and keep them open for use entirely at his own expense every day from daybreak to the seventh hour for women, and from the eighth hour to the second hour in the evening for men, at the discretion of the procurator in charge of the mines.

ii. He shall be required to provide a proper supply of running water for the heated rooms, to the bath tub up to the highest level and to the basin, for women and well as for men.

iii. The lessee shall charge men one half an as each and women one as each.

iv. Imperial freedmen and slaves in the service of the procurator or on his payroll are to be admitted free, as are children and soldiers.

v. At the expiry of the lease the lessee, or his partner or agent, shall be required to return in good condition all the bath equipment consigned to him, excepting any rendered unusable through age.

vi. He shall be duly required to wash, dry and coat with fresh grease every thirty days the bronze implements which he uses.

vii. If any needed repair prevents the proper operation of the baths, the lessee shall be entitled to prorate the rental for that period.

viii. Excepting this, whatever else he may do for the purpose of operating the said baths, he shall be entitled to no reduction in rental.

ix. The lessee shall not be allowed to sell wood except for branch trimmings unsuited for fuel.

x. If he does anything in violation of this, he shall have to pay to the fisc 100 sesterces for each sale.

xi. If these baths are not properly kept open for use, then the procurator of the mines shall have the right to fine the lessee up to 200 sesterces every time they are not kept open properly.

xii. The lessee shall at all times have on hand a supply of wood sufficient for .... days.

4. OF THE SHOEMAKING TRADE

i. Anyone who makes any of the shoes or thongs which shoemakers customarily handle, or who drives or sells shoemaker's nails, or who is convicted of selling within the mining district anything else which shoemakers are entitled to sell shall have to pay double to the lessee, or to his partner or agent.

ii. The lessee shall sell nails in accordance with the regulations of the iron mines.

iii. The lessee, or his partner or agent, shall have the right to obtain security.

iv. No one will be allowed to repair shoes, except to mend or repair his own or his master's [?].

v. The lessee shall be required to offer all types of shoes for sale; if he does not, everyone shall have a legal right to purchase wherever he wishes.

5. OF THE BARBERING TRADE

i. The lessee shall be entitled to operate with the assurance that no one else in the village of the Vipasca mines or within the mining district thereof shall practise barbering for profit.

ii. Anyone who does so practise barbering shall have to pay to the lessee, or his partner or agent, ... denarii for each use of razors, and the said razors shall be forfeited to the lessee.

iii. Slaves attending to their masters or their fellow slaves are excepted.

iv. Itinerant barbers not sent out by the lessee shall not have the right to practise barbering.

v. The lessee, or his partner or agent, shall have the right to obtain security.

vi. Anyone who hinders his obtaining security shall have to pay five denarii for each such act of hindrance.

vii. The lessee shall engage one or more skilled workers in proportion [to the need].

6. OF THE FULLERS' SHOPS

i. No one to whom the lessee, or his partner or agent, has not leased or granted the right shall be allowed to do work on new or used clothing for pay.

ii. Anyone convicted of any violation of this shall have to pay the lessee, or his partner or agent, three denarii for each garment.

iii. The lessee, or his partner or agent, shall have the right to obtain security.

7. OF THE FEE FOR EXPLOITING SLAG HEAPS AND ROCK DEPOSITS

i. Anyone who wishes in the district of the mines of Vipasca ..... to clean, process, break up, sift or wash silver or copper slag or slag dust scooped out by measure or weight, or who undertakes to execute work of any kind in the quarries, shall within the next three days declare what slaves and hired labourers he will send to do this, and shall pay the lessee .... each month on or before the last day of the month. If he does not do so, he shall have to pay double.

ii. Anyone who brings into the mining district scoopings of copper or silver from ore dumps in other places shall have to pay one denarius per 100 pounds to the lessee, or to his partner or agent.

iii. If anything is owed to the lessee, or to his partner or agent, in accordance with this section of the regulations, and is not paid or secured on the day on which the obligation is incurred, he shall have to pay double.

iv. The lessee, or his partner or agent, shall have the right to obtain security, and any of the said slag that is cleaned, processed, broken up, sifted, or washed and any stone blocks that are dressed in the quarries, shall be forfeited to the lessee, or his partner or agent, if whatever is owed to him is not paid.

v. Slaves and freedmen of silver and copper smelters working in the smelting-plants of their masters or patrons are excepted.

8. SCHOOLTEACHERS

It is decreed that schoolteachers are exempt from taxation at the hands of the procurator of the mines.

9. CLAIMS TO WORKINGS, OR PERMIT FEE [?]

Anyone who, in accordance with the regulations issued for the mines, stakes a claim to or occupies a mineshaft or the site of a mineshaft within the district of the mines of Vipasca with a view to maintaining a legal right shall within the next two days after such claim or occupancy declare to the lessee of the this revenue, or his partner or agent .... [THE REST IS LOST]

NOTE : This translation is based on that which appears in N. Lewis & M. Reinhold, Roman Civilization Sourcebook II : The Empire, New York, 1966, 191-94.

VIPASCA B = LEX METALLIS DICTA

(= *FIRA* I, 104)

N.B. The headings to each section are my own, and are not contained in the Latin text, but may be useful as points of reference.

PREAMBLE

....to his dear Ulpius Aelianus, greetings.

PAYMENT FOR SMELTING OF [COPPER ?] ORE

1. i. ...... he shall make immediate payment.

ii. If anyone does not do so and is convicted of having smelted ore before paying the price as specified above, his share as occupier shall be confiscated and the entire mineshaft (*puteus*) shall be sold by the procurator of the mines.

iii. Anyone who proves that a tenant (*colonus*) has smelted ore before paying the price of the half share belonging to the fisc shall receive one quarter (of this sum).

THE LEASING OF SILVER WORKINGS

2. Silver shafts must be worked in accordance with the details contained in these regulations. Prices for them will be kept in accordance with the liberality of the most sacred Emperor Hadrian Augustus, whereby the ownership of the share belonging to the fisc belongs to

the first person to offer the price for the working and pay down to the fisc the sum of 4,000 sesterces.

REGULATIONS TO ENSURE PRODUCTION IN SILVER WORKINGS

3. i. If anyone strikes ore in one out of five mineshafts, he shall, as stated above, carry out the work in the others without interruption.

ii. If he does not do so, another shall have the legal right to take possession.

4. If anyone after the twenty-five day period granted for raising working capital actually begins regular operations, but then stops operations for ten consecutive days, another shall have the right to take possession.

5. If a mineshaft sold by the fisc lies idle for six consecutive months, another shall have the legal right to take possession on condition that when the ore is extracted therefrom, one half shall be, according to customary practice, reserved for the fisc.

THE RIGHT TO FORM ASSOCIATIONS (SOCIETATES) IN SILVER WORKINGS

6. i. The occupier of mineshafts shall have the right to have any partners (socii) he wishes on condition that each partner contribute his proportionate share of the expenses.

ii. If anyone does not do so, then the one who covers the expenses shall have an account of the expenses covered by himself posted on three consecutive days in the most frequented place in the forum, and shall demand of his partners through the public crier that each contribute his proportionate share of the expenses.

iii. If anyone does not contribute or does anything with malice aforethought to avoid contributing or to deceive one or more of his partners, he shall not have his share of the mineshaft, and that share shall belong to the partner or partners who cover the expenses.

7. Tenants who cover expenses in such mineshafts in which there are several partners shall have a legal right to recover from their partners anything that is shown to have been expended in good faith.

8. i. Tenants shall also have the right to sell among themselves, at as high a price as they can, shares of mineshafts purchased from the fisc and paid for.

ii. Anyone who wishes to sell or buy a share shall submit a declaration to the procurator who is in charge of the mines; otherwise he shall not be allowed to buy or sell.

iii. Anyone who is a debtor to the fisc shall not be allowed to give away his share.

SMELTING AND STEALING OF ORE

9. i. Ore extracted and lying near the shafts must be conveyed to the smelters by those to whom it belongs between the hours of sunrise and sunset.

ii. Anyone convicted of having removed ore from the shafts after sunset or by night shall have to pay to the fisc the sum of 1,000 sesterces.

10. i. A stealer of ore, if he is a slave, shall be whipped by the procurator and sold on this condition, that he shall be kept perpetually in chains and not tarry in any mine or mining district; the price of the slave shall belong to the owner.

ii. If the thief is a free man, the procurator shall confiscate his property and debar him forever from the mining district.

THE SAFETY OF MINING SHAFTS : THE MAINTENANCE OF PROPS

11. All mineshafts shall be carefully propped and reinforced, and the tenant of each working shall provide new and suitable replacements for rotten materials.

12. It shall be forbidden to touch or damage pillars or props left for reinforcement, or to do anything with malice aforethought to render the said pillars or props unsafe.

13. If anyone is convicted of having injured, weakened, or damaged [?] a mineshaft, or of having done anything with malice aforethought to render such mineshafts unsafe, if he is a slave, he shall be whipped at the discretion of the procurator and sold by his master on this condition, that he shall not tarry in any mine [or mining district]; if he is a free man, the procurator shall confiscate his property and debar him forever from the mining district.

PROSPECTING AND THE MAINTENANCE OF DRAINAGE DITCHES IN COPPER WORKINGS

14. Anyone who operates copper shafts shall avoid the ditch (<u>cuniculum</u>) that drains the water from the mines and leave a space of not less than fifteen feet on either side.

15. i. It shall be forbidden to damage the ditch.

ii. The procurator shall permit the driving of a drift (ternagus) from the ditch for the purpose of discovering new mineral, on condition that the drift be not more than four feet in width and depth.

16. It shall be forbidden to look for or extract ore within fifteen feet on either side of the ditch.

17. If anyone is convicted of having done anything contrary to the regulations in the drifts, if he is a slave, he shall be whipped at the discretion of the procurator and sold by his master on this condition, that he shall not tarry in any mine [or mining district]; if he is a free man, the procurator shall confiscate his property and debar him forever from the mining district.

PROSPECTING AND THE MAINTENANCE OF DRAINAGE DITCHES IN SILVER WORKINGS

18. Anyone who operates silver shafts shall avoid the ditch that drains the water from the mines and shall leave a space of not less than sixty feet on either side. He shall exploit those shafts occupied by him or allotted to him within the boundaries prescribed, and shall not advance beyond them nor collect leavings (egbolae) nor drive drifts outside the limits of the mineshafts allotted. ....

NOTE : This translation is closely based on that in N. Lewis & M. Reinhold, Roman Civilization Sourcebook II : The Empire, New York, 1966, 188-90.

APPENDIX III. ROMAN CETARIAE IN LUSITANIA

A. THE ALGARVE

C 01 ILHEU DE BALEEIRA, Forte de Beliche, Ponta da Sagres

Traces of many Roman constructions reported, as well as many fragments of Roman pottery. The location of these constructions (right on the coast on what is now an island) suggest some connexion with fishing. Possible Roman amphora kiln in vicinity.

Ref. : AP, 15, 1910, 211.
Arq. de Beja, 23-24, 1966-1967, 128.
AP, iii, 5, 1971, 159 (kiln ?).

C 02 SALEMA

Foundations of tanks lined with cocciopesto. Three bronze fish-hooks and mosaic. Punic pottery perhaps from this site.

Ref. : AP, 15, 1910, 211.
ARA, I, 77.

C 03 BURGAU

Roman tanks reported, together with fish-hooks. Mosaic and numerous Roman coins also found, many of the Late Empire, suggesting occupation until the end of C4/start C5 A.D.

Ref. : AP, 15, 1910, 218.

ARA, I, 107.

C 04 SENHORA DA LUZ

Roman tanks lined with cocciopesto extending for 160m. Four copper fish-hooks. Many amphora fragments and "baths" reported. Again the latter would better be viewed as as part of the industrial complex. Coin of Theodosius points to occupation in C4.

Ref. : Bull. hisp. 8, 1906, 115.
AP, 15, 1910, 218-9 & fig. 5.
ARA, I, 108-111.

C 05 PAUL, LAGOS

Remains of Roman tank (0.30m deep), lined with cocciopesto. Fragments of amphorae now deposited in museum in Lagos.

Ref. : AP, 15, 1910, 221.
Arq. de Beja, 23-24, 1966-1967, 128.
ARA, I, 118 & fig. 39 (amphora).

C 06 BOCA DO RIO

Four tanks preserved with cocciopesto lining. The now liquidated Companhia de Pescarias do Algarve also built warehouses on top of more Roman tanks. Baths with hypocaust also reported, but this would more plausibly be explained as part of the industrial complex. Five copper and four bronze fish-hooks also found and a small amphora, now preserved in the Museu Regional de Lagos. Recent excavations have revealed mosaics and various nuclei of *cetariae*, suggesting that this was a very large establishment.

Ref. : Bull.hisp., 8, 1906, 113-114.
AP, 15, 1910, 212-4 (plan = fig. 3).
AEA, 26, 1953, 120 (amphora).
ARA, I, 80, 92-93 & fig. 7 (p. 96).
Informação arq., 5, 1982-3, 148-149.

C 07 VAU (3 km SSW of Mexilhoeira)

Various Roman tanks (1.50m long x 1.03m wide x 1.85m deep), lined with cocciopesto, found near the mouth of the river Alvor. Remains of fish-bones in some of the better preserved tanks.

Ref. : Bull.hisp., 8, 1906, 112 & fig. 2 (p. 115).
AP, 15, 1910, 226-227.
ARA, I, 124-125.
Museus de Portugal, 1, 1978, 43-49.

C 08 MEXILHOEIRA GRANDE, ABICADA

Possible Roman tanks found 20m from Roman villa which was occupied from C1 - C4 A.D. Amphora (Almagro 50) also found.

Ref. : J.-G. Gorges, Les villas hispano-romaines, sv. PS 38.

C 09 PORTIMÕES, PORTIMÃO

Series of tanks lined with cocciopesto discovered on banks of river Portimões. Amphora bases, three bronze fish-hooks, terracotta net weights and fragments of tegulae and imbrices (with potter's marks) also found. Roman baths also reported.

Ref. : AMA, II, 569-570.
Bull. hisp., 8, 1906, 116-117 & fig. 3 (plan).
AP, 15, 1910, 228.

ARA, I, 128-131 & figs. 43 (plan), 44 & 45 (photos).
AP, iii, 7-9, 1974-1977, 246-248.
Museus de Portugal, 1, 1978, 49-57 (photo, map & plan).

C 10 FERRAGUDO, CASTELO DE ARADE

Three Roman tanks (approx. 1m x 1m x 1m), covered with mortar reported. Coin of Nero. Mosaics and some stucco found in locality, suggesting possible Roman villa.

Ref. : E. da Veiga, Inventario do Museu Arqueologico do Algarve, 1883, 40.
ARA, I, 135-136.

C 11 ARMAÇÃO DE PERA

Tanks submerged beneath the sea reported at end of C19. Numerous fragments of Roman ceramics found on beach.

Ref. : AMA, II, 368.
Bull. hisp., 8, 1906, 117.
ARA, I, 141.

C 12 CERRO DA VILA

Two tanks, lined with cocciopesto, found closely annexed to Roman villa site.

Ref. : ARA, I, 143-144.
AP, iii, 5, 1971, 201-214 (with plan).
AP, iii, 6, 1972, 251-262 (with plan).
J.-G. Gorges, Les villas hispano-romaines, sv. PS 41.

C 13 QUARTEIRA

Traces of Roman tanks reported 10m beneath the sea. Local tradition also recalls Roman cetaria. Two bronze fish-hooks and amphora found nearby.

Ref. : Bull. hisp., 8, 1906, 117.
Arqueologia e Historia, viii, 12, 1966, 76-79.
ARA, I, 149-150.

C 14 LOULÉ VELHO

Two tanks lined with cocciopesto reported with connecting limestone wall. Coin finds suggest an occupation at least from C1 - C4 A.D.

Ref. : Arq. e Historia, viii, 12, 1966, 75-76 & figs. 17 & 18 (photos);
ARA, I, 151-152 & figs. 56-61 (photos).

C 15 OSSONOBA (FARO)

(A) LARGO DA SÉ

Roman tanks, lined with cocciopesto, found in the excavations of the Largo da Sé. Four copper fish-hooks and the remains of fish-bones and shell-fish shells found in vicinity, as well as Roman ceramics.

Ref. : ARA, I, 186-199, 201 & figs. 62 & 62a.

(B) LARGO DA MADALENA

Possible cetaria in zone of Largo da Madalena near Rua Conseilheiro Bivar. Six amphorae found standing vertically with traces of salt, fish and fish-sauce. Coin of Nero also found.

Ref. : <u>Anais do Municipio de Faro</u>, 6, 1976, 157.
<u>Anais do Municipio de Faro</u>, 14, 1984, 152.

(C) AVENIDA DA REPUBLICA/TRAVESSA DA MADALENA

Three Roman salting-tanks located, but not excavated. Roman column also found in vicinity.

Ref. : <u>Anais do Municipio de Faro</u>, 14, 1984, 153.

C 16 OLHÃO

Series of Roman tanks with cocciopesto revetment found in 1950. Five bronze fish-hooks and numerous garum amphorae reported in the vicinity of Roman villa site. Occupation from C1 - start of C5 A.D.

Ref. : <u>Industria Portuguesa</u>, 273, Nov. 1950.
<u>Arq. de Beja</u>, 23-24, 1966-1967, 128.
<u>ARA</u>, I, 215-216 & figs. 78 (plan) & 79 (fish hooks).
J.-G. Gorges, <u>Les villas hispano-romaines</u>, sv. PS 39.

C 17 ALFANXIA, Olhão

Remains of walls and cocciopesto-lined tank (1.80m x 1.60m) reported in the vicinity of Roman amphora kiln.

Ref. : J. Fernandes Mascarenhas, <u>Por Terras do Algarve</u>, 1974, 13.

C 18 BALSA : Quinta da Torre de Ares

Tank reported (4m x 2.80m x 1.20m), lined with cocciopesto and with its corners rounded, in area of substantial Roman remains. Seventeen fish-hooks and further fragments of hooks now preserved in the Museu Nacional de Arqueologia e Etnologia in Lisbon, with ten more hooks in private collection. Two bronze needles for net-making were also found here. Numerous amphorae (some complete, but mostly fragments) and tegulae and imbrices also found on the site.

Ref. : Bull. hisp. 8, 1906, 118.
ARA, I, 232-233, 255 & figs. 82 (plan) and 109 (fish-hooks).
AP, iii, 7-9, 1974-1977, 250.

C 19 BALSA : Quinta das Antas

Group of tanks, lined with cocciopesto, were discovered on the left bank of the Ribeira das Antas, some partly submerged. Two fish-hooks and many amphorae also found on the site.

Ref. : Bull. hisp., 8, 1906, 118.
ARA, I, 288-290, 296 & figs. 115 & 116.

C 20 CACELA

Three tanks, all lined with cocciopesto, reported in vicinity of Roman villa. One bronze fish-hook and many amphorae (complete and fragmented) now in Museu Nacional in Lisbon.

Ref. : Bull. hisp., 8, 1906, 118-119.
AP, 24, 1919-1920, 230 & fig. 136.
ARA, I, 306-307.
J.-G. Gorges, Les villas hispano-romaines, sv. PS 35.

C 21 QUINTA DO MURO

Four Roman tanks discovered here, as well as a single fish-hook.

Ref. : *Bull. hisp.*, 8, 1906, 119-120 & fig. 6.
ARA, I, 309-311.

C 22 QUINTA DO LAGO, S. BARTOLOMEU DE CASTRO MARIM

Report of Roman *cetariae*, with abundant ceramic material from C3 & C4 A.D. (late *terra sigillata hispanica*, *terra sigillata clara*) and "*amphorae of this era*" : ? Lusitanian Garum II, III, IV. Later Arab settlement on site.

Ref. : V. Gonçalves, *Clio/Arq.*, 1, 1983-84, 270-271 & esp. 280-281.

B. SADO/TAGUS BASIN

C 23 ATOUGUIA, PENICHE

Possible Roman fish-processing tanks (2 complete and 1 ruined), 3m x 2m, revetted in cocciopesto reported in area of noted salt production.

Ref. : *Ethnos*, 5, 1966, 127-128 & photo (p. 130, but not very revealing).

C 24 CASAIS VELHOS, OUTEIRO DAS VINHAS, GUINCHO (Cascais)

A Roman tank (9m x 3.65m) reported from elevated position with connecting reservoir and canalizations. A little to the west, two smaller tanks (1.20m x 0.62m and 0.77m x 0.61m) revetted in cocciopesto and with rounded corners. Roman coarseware and terra sigillata pottery, lead weights, columns and coins of Constantius, Magnentius, Gratian, Magnus Maximus and Honorius also found.

Ref. : F.J. do Amaral Figueiredo & A. do Paço, V Congr. Arq. del Sudeste esp. & I Congreso Nacional de Arqueologia, Almeria, 1949 (no pg. nos.).
Arq. de Beja, 23-24, 1966-1967, 128.
AP, iii, 2, 1968, 208.

C 25 CASILHAS

Four cetariae unearthed in the Largo Alfredo Dinis in December 1981. Various fragments of terra sigillata (Italian and South Gaulish), thin-walled ware, handles and rims of amphorae (1 Lusitanian Garum I identified), various terracotta net weights. Material suggests that site in operation from the end of Cl B.C.

Ref. : O Diario, 17.xii.81, 1, 5.
L. Barros & C. Amaro, Almadan, 4-5, 1984-1985, 33-4.

C 26 COURELA DO MARCO, GARROCHEIRA, conc. Benavente

Possible Roman cetaria in close proximity to Roman amphora kiln in valley of river Sorraia.

Ref. : C. Amaro, Arqueologia, 3, Jun. 1981, 131-132.

C 27 OLISIPO (= LISBON)

(A) CASA DOS BICOS

Four *cetariae* and other structures revealed on right bank of Tagus. Fragments of Roman stucco, mosaic *tesselae*, late Campanian ware and *terra sigillata* pottery (unspecified type), fragments of late Roman amphora (lip) and great quantities of coarseware, bones and shells. Possible Roman quay located in vicinity.

Ref. : *Informação arq.*, 4, 1981, 154.
Informação arq., 5, 1982-83, 149.

C 28 ALFARIM, CABO ESPICHEL

Roman *cetaria* reported.

Ref. : *Arq. de Beja*, 23-24, 1966-1967, 128.

C 29 CREIRO

Roman *cetariae* reported.

Ref. : *AP*, 10, 1905, 187 & fig. 181 (p. 188).
AP, 27, 1925-1926, 174.
J. Marques da Costa, *Novos elementos para a localização de Cetobriga*, Setubal, 1960, 16.

C 30 MOUGELAS, RASCA

Roman tanks, lined with cocciopesto, discovered beneath the sea, containing various fragments of molluscs, including Purpura haemostoma, Murex trunculus and Murex brandarix.

Ref. : AP, 10, 1905, 189 & fig. 181.
AP, 27, 1925-1926, 175-176.
AP, iii, 4, 1970, 230.

C 31 QUINTA DA COMENDA

Ruined tanks, lined with cocciopesto, of similar type and dimensions to those at Troia (C 37), now submerged beneath the water level. Bronze coin of Constantius found in one tank. Canalized water supply from the Ribeira da Ajuda. Fragments of amphorae, tegulae, imbrices and various coins of the C4 A.D. Great quantity of fish-bones and molluscs. Excavations of 1978 revealed Roman material dated from the start of C1 A.D. to the end C4.

Ref. : AP, 10, 1905, 187-191 & fig, 181.
AP, 27, 1925-1926, 174-175.
AP, iii, 4, 1970, 230.
Informação arq., 1, 1977-1978, 32.

C 32 SETÚBAL (? CAETOBRIGA)

(A) PRAÇA DO BOCAGE

Five ruined Roman tanks with cocciopesto lining. From material contained within them they appear to have been used from the end of C1 A.D. until C3, after when they were used as rubbish dumps.

Ref. : J. Marques da Costa, op.cit (sv. C 29).
C. Tavares da Silva, Escavações arqueologicas na Praça do Bocage : 2000 anos de historia, Setubal, 1980.
C. Tavares da Silva & A. Coelho-Soares, Setúbal arq., 6-7, 1980-81, 249-294.

(B) TRAVESSA FREI GASPAR

Double file of tanks of varying dimensions : 0.80m x 1.20m (the six smallest), 1.60m x 3m (the largest), 2.20m x 2.20m (the rest). Their depth varies between 1.30m and 1.35m. All lined with cocciopesto and with their corners rounded. Remains of fish-bones (sardines and possibly tunny). Some of the tanks had a false bottom, suggesting a first phase and then later reoccupation. Early material includes South Gaulish terra sigillata, thin-walled ware (2nd half C1 A.D.); later material terra sigillata clara (Hayes 73) and late terracotta lamps (type Dressel 31), suggesting occupation into the C5. Fragments of Beltran II & Lusitanian Garum I, II and IV amphorae.

Ref. : L. Ferrer Dias in Descobertas arq. do sul de Portugal, Lisbon/Setúbal, 1980, 57-58.
Informação arq., 2, 1979, 82-3 (with plan).

(C) RUA ANTONIO JANUARIO DA SILVA

Report of Roman cetaria complex.

Ref. : C. Tavares da Silva & J. Soares, Almadan, 3, 1984, 2.

C 33 CACHOFARRA

Roman fish-processing tanks reported on the very bank of the Sado.

Ref. : AP, 10, 1905, 187.

C 34 PEDRA FURADA

Roman tanks reported.

Ref. : AP, 10, 1905, 187.
AP, 12, 1907, fig. 274a.

C 35 SENHORA DA GRAÇA

Roman tanks reported.

Ref. : AP, 10, 1905, 187.
AP, 12, 1907, fig. 274a.

C 36 SANTA CATHARINA

Roman cetariae reported at two locations : Moinho Novo and Ponta de Areia.

Ref. : AP, 10, 1905, 187.

C 37 TRÓIA

Most extensive Roman fish-processing establishment yet discovered in Lusitania. The site, on a thin spur of land jutting into the Sado estuary, consists of a large number (estimated at 2000 - 3000 by a C19 traveller) of tanks of varying dimensions, all revetted with cocciopesto. The tanks are grouped into roughly 52 separate units of production. Some of these units were covered with a roof. The area of the greatest concentration of tanks extends for 1 km, but the whole site stretches for almost 4 km. In addition to the tanks, a columbarium has been uncovered with niches in its interior walls for funerary urns, as well as baths and hypocausts. Again such hypocausts may have had some connexion with the industrial processes. Various wells and cisterns on the site provided the water needed for the production of fish-sauces (as at Cotta in north Africa). A early Christian baptistry has also been found, with its interior walls stuccoed and frescoed. A series of elegant town houses with mosaics have been found in the Rua da Princesa. Scattered along the beach are innumerable fragments of <u>tegulae</u>, <u>imbrices</u>, <u>terra sigillata</u> (Arretine, South Gaulish, Spanish and African), coarseware and especially amphorae of many different types. Fish-hooks, needles for net-making and net weights all emphasise the importance of fishing to the establishment.

Various inscriptions and reliefs have come to light, most notably a relief in honour of Mithras. In addition to the <u>columbarium</u>, other necropoleis have been located. As for chronology, there is no evidence for the occupation of the site before the end of C1 B.C.. There seems to have been some abandonment in C4 A.D., but the site continued to be occupied until C6, since Late Roman C wares have been found, dating from the start of C6.

Ref. : A.I. Marques da Costa, <u>AP</u>, 4, 1898, 344-352.
<u>AP</u>, 26, 1923-1924, 314-328.
<u>AP</u>, 27, 1925-1926, 165-181.
<u>AP</u>, 29, 1930-1931, 2-31.
<u>Arq. de Beja</u>, 23-24, 1966-1967, 125-126, 128.
J. Soares, <u>Estação romana de Troia</u>, Setúbal, 1980.

C 38 SALACIA (= ALCACER DO SAL)

Report of Roman tanks in the lower Sado estuary.

Ref. : V. Correia, <u>Obras IV</u>, Coimbra, 1972, 132.

C 39 SINES

Report of Roman fish-processing tanks on Atlantic coast near Sines. Numerous fish-hooks and net weights, and many fragments of amphorae, displayed in Museu Municipal de Sines.

Ref. : A. Soledade, Sines : terra de Vasco da Gama (2nd ed.), Sines, 1981, 16-17 & fig. 4.

C 40 ILHA DO PESSEGUIERO (off Sines)

Two cetariae discovered in 1982 and 1983.

(A) A walled complex 8.6m x 7.8m, consisting of a rectangular area floored with cocciopesto, surrounded by 8 tanks of varying dimensions (the smallest 1.3m x 1.2m, the largest 2.2m x 2m), dug into the bedrock to a depth of 1.3 - 2m.

(B) A walled complex 17m x 6.8m, consisting of a central area floored with cocciopesto, surrounded on three sides by 10 tanks of varying dimensions (some square : 1.30m x 1.30m, some rectangular : one 2.2m x 2.4m, others 1.3m x 2.5m). To the extreme north of the complex 2 square areas for preparation of the fish, as well as storerooms.

From pottery finds (terra sigillata hispanica from Andujar, Baetica; terra sigillata clara A, C and D) the complexes were built in the first half of the C2 A.D., and continued in operation until the mid C4 A.D. Amphorae finds include fragments of Lusitanian Garum II, III & IV. In addition fish-bones and mollusc shells of various species (Patella, Myrtilus, Monodonta lineata, Thais haemastoma), as well as bronze fish-hooks and bone needles for net-making found in complexes.

Ref. : C. Tavares da Silva, J. Soares & L. Ferrer Dias, Setúbal Arq., 6-7, 1980-81, 219-247.
C. Tavares da Silva, Almadan, 2, 1983-84, 20-22.

APPENDIX IV. SOME TEXTS ON ROMAN FISH-PROCESSING

(A) GEOPONIKA XX.46 : The preparation of fish-sauces

1. Liquamen is produced in the following way. The insides of fish are thrown into a tank and salted. Small fish - especially smelts, small mullets, sprats, anchovies or anything else that seems to be small - are all salted in the same way; and by macerating them in an enclosed space they are pickled in the sun.

2. And when the pickling has been carried out in the heat, garum is produced from this briny solution in the following way. A large, strong basket is placed in the middle of the tank containing the aforementioned fish and the garum flows into the basket; in the same way the so-called liquamen is strained off through the basket and removed. The remaining solution is hallec.

3. In Bithynia it is prepared as follows. Best of all you take small or large sprats (but anchovies, horse mackerel, mackerel or even hallec will do) and mix them all up together and place them in a baking trough, usually used for kneading dough, and add two Italic _sextarii_ of salt to one _modius_ of fish and mix it up well, so that it blends together with the salt. After leaving it for one night, put it in an earthenware jar and place it without its lid in the sun for two or three months, stirring it with a stick at intervals; then take it and seal it with lids.

4. Some people add up to a _sextarius_ of old wine to every two _sextarii_ of fish.

5. But if you want to use the garum immediately, do not leave it to bake in the sun, but boil it. You can make it as follows : after testing the briny liquid, by throwing in an egg and seeing if it will float (if it sinks, there is insufficient salt), then throw the fish into some brine in another earthenware pot, put in some oregano and then heat it up until it boils, that is until it starts to separate out a little. Some people add mustum boiled down to a third part. Then when it has cooled, put it in a colander and strain it two or three times in the colander, until it comes out free of any impurities and then seal it with lids.

6. The finest garum, which is called "blood-sauce" (haimation), is produced in the following way : take the insides of tunny fish with the gills, serum and the blood and mix them with as much salt as is necessary. Leave it in the vessel and, after two months on average you break through the vessel and the so-called "blood-sauce" garum pours out.

TEXT = Geoponica ed. Beckh (Teubner Texts), pp. 528-9.

(B) MANILIUS, ASTRONOMICA, BOOK FIVE

(a) Choice garum and hallec (lines 667-75)

Furthermore, when the victims lie dead along the shore, a second slaughter takes place on top of the first. The fish are torn to shreds and a single body is divided to serve separate ends. One part is better if its juices are surrendered, another if they are retained. In one case a valuable fluid is discharged, which yields the choicest part of the blood : flavoured with salt, its imparts a relish to the palate (i.e. haemation, "blood-sauce" garum). In the other case all the pieces of the decaying carcase are blended together and merge their shapes until every distinguishing feature has been lost : they provide food with a condiment for general use (i.e. hallec).

(b) Liquamen (lines 676-81)

.... Or .... a shoal of scaly creatures .... are surrounded and drawn from the water by a huge drag net and fill large tanks and vats; their common discharge of liquid is exuded upon each other, for their inwards melt and issue forth as a stream of composition (i.e. liquamen).

(c) Salt (lines 682-89)

.... they prepare a wide stretch of hardened ground and surround it with firm walls; next they direct on to this waters channelled from the nearby sea and then deny them exit by closing sluice-gates. So the floor holds in the waves and begins to glisten as the water is drained off by the sun. When the sea's dry element (i.e. salt) has collected, Ocean's white locks are shorn for use at table and huge mounds are made of the solid foam.

NOTE : This translation is based on that of A.E. Housman, Loeb Classical Library ed. of Manilius.

APPENDIX V. AMPHORA KILNS IN LUSITANIA

NOTE ON ABBREVIATIONS

Where possible, some very rough indication is given of of the relative proportions of types of amphora etc. produced (viz. *, **, or ***).

Abbreviations of amphora types etc. are given under the first site, K 01.

Supplementary bibliography is given at the end of the Appendix for those kilns marked + .

A. SADO ESTUARY

K 01 QUINTA DA ALEGRIA

Lusitanian Garum I	**	(= LG I)
Lusitanian Garum II	*	(= LG II)
Lusitanian Garum III	*	(= LG III)
Lusitanian Garum IV	***	(= LG IV)
Coarseware	*	(= CC)

Ref. : A. Coelho-Soares & C. Tavares da Silva, <u>Set. Arq.</u> 5, 1979, 205-21.

K 02 HERDADE DO PINHEIRO +

LG I	*
LG II	*
LG III	*
LG IV	***
CC	***

Ref. : D. F. de Almeida, G. Zbyszewski & O. da V. Ferreira, <u>AP</u>, iii, 1971, 155-65.

K 03 ABUL

LG I ***
CC ***

Ref. : A.M. Dias Diogo et al., Série Arqueologica, Museu do Mar, Cascais, I.4, 1985.

K 04 MONTE DE ENCHURRASQUEIRA I

LG I ***
LG IV *
CC **

Ref. : A.M. Dias Diogo, Conimbriga, 22, 1983, 209-15; A.M. Dias Diogo et al., Serie Arq., Museu do Mar, Cascais, I.3, 1984.

K 05 MONTE DE ENCHURRASQUEIRA II +

LG I ***
CC **

Ref. : unpubl.

K 06 VALE DE CEPA

LG I ***
LG IV ***
CC **

Ref. : A.M. Dias Diogo, Conimbriga, 22, 1983, 212-4.

K 07 MONTE DO BUGIO

LG I ***
CC **

Ref. : A.M. Dias Diogo, Conimbriga, 19, 1980, 147-50.

K 08 MONTE DO BUGIO II +

LG I ***

Ref. : unpubl.

K 09 HERDADE DO BARROSINHA I +

LG I ***
CC **

Ref. : unpubl.

K 10 HERDADE DO BARROSINHA II +

LG I ***
CC **

Ref. : unpubl.

K 11 ALCÁCER DO SAL +

LG I ***
CC **

Ref. : unpubl.

K 12 HERDADE DA BATALHA (small kiln) +

LG I *
CC *

Ref. : unpubl.

B. TAGUS ESTUARY

a) Left Bank

K 13 COURELA DO MARCO, GARROCHEIRA, BENAVENTE

LG I	***
CC	*

Ref. : C. Amaro, Arqueologia, 3, 1981, 131-2.

K 14 HERDADE DO RIO FRIO, ALCOCHETE

unspecified

Ref. : L. Barros & C. Amaro, Almadan, 4-5, 1984-85, 33-34.

b) Right Bank

K 15 CARROIOS

unspecified

Ref. : L. Barros & C. Amaro, loc. cit.

C. THE ALGARVE

K 16 PONTE DE SAGRES

unspecified

Ref. : D.F. de Almeida, AP, iii, 5, 1971, 159.

K 17 PRAIA DO MURTINHAL

unspecified

Ref. : E. da Veiga, AP, 15, 1910, 211.

K 18 LAGOS

local type of amphora

Ref. : A.J. Parker, Méthodes classiques et méthodes formelles dans l'étude des amphores, Paris/Rome, 1977, 39.

K 19 SAO JOÃO DA VENDA

local type of amphora

Ref. : ARA, II, 166-70.

K 20 ALFANXIA

local type of amphora

Ref. : J. Mascarenhas, Por Terras do Algarve, 1974, 11-12.

K 21 QUINTA DA TORRE DE ARES (BALSA)

LG II	*
? LG IV	*
CC	**

Ref. : ARA, I, 232-33.

K 22 VALE DO BOTO, SÃO BARTOLOMEU DE CASTRO MARIM

LG IV	**
local type	***

Ref. : V. Gonçalves et al., Clio, 2, 1980, 71-79.

K 23 SÃO BARTOLOMEU DE CASTRO MARIM

LG I	*
LG II	*
LG IV	*
CC	**

Ref. : M. Maia, Clio, 1, 1979, 141-48; J.L. de Vasconcelos, AP, 4, 1898, 329-36.

NOTE : I am most grateful to Dr. J.C. Faria of the Museu Municipal de Alcacer do Sal & Dr. F. Reiner & A. Carvalho of the Museu do Mar, Cascais for much information on amphora kilns (many as yet unpublished). I give below the titles of forthcoming articles on certain kilns :

K 02 : A.M. Dias Diogo et al., 'Material dos fornos romanos do Pinheiro no Museu do Mar, Cascais' in Serie Arqueologica do Museu do Mar (forthcoming)

K 05 : A.M. Dias Diogo et al., 'Fornos romanos da Enchurrasqueira II'

K 08 : J.C. Faria, M. Ferreira & A.M. Dias Diogo, 'Fornos romanos do Bugio II'

K 09 : A.M. Dias Diogo & J.C. Faria, 'Fornos romanos da Barrosinha I'

K 10 : J.C. Faria & A.M. Dias Diogo, 'Fornos romanos da Barrosinha II'

K 11 : A.M. Dias Diogo & J.C. Faria, 'Fornos romanos de Alcácer do Sal';

A.M. Dias Diogo, J.C. Faria & M. Ferreira, 'Material anforico do lado poente do Castelo de Alcácer do Sal'

K 12 : A.M. Dias Diogo et al., 'Fornos do Monte da Batalha'

APPENDIX VI. LUSITANIAN AMPHORAE

NOTE ON ABBREVIATIONS

M. Beltran Lloris, Las anforas = Las anforas romanas de España (= Monografias arqueologicas, VIII), Zaragoza, 1970.

A.J. Parker, 'Lusitanian amphoras' = Méthodes classiques et méthodes formelles dans l'étude des amphores romaines (= Coll. de l'Ecole française de Rome, 32), Paris/Rome, 1977, 35-46.

S.J. Keay, LRA = Late Roman Amphorae in the Western Mediterranean : a typology and economic study : the Catalan evidence, 2 vols., (B.A.R. International Series, 196), Oxford, 1984.

N.B. The numbers in brackets give some rough orientation as to the quantity of amphorae found.

A. LUSITANIAN GARUM I (= BELTRAN IVb/DRESSEL 14)

1. LUSITANIA

São Bartolomeu de Castro Marim (20) KILN

Ref. : M. Maia, Clio, 1, 1979, 141-148.

Quinta da Alegria (Setubal) (28 + frags.) KILN

Ref. : A. Coelho-Soares & C. Tavares da Silva, Setúbal Arq., 5, 1979, 205-221.

Tróia (3 + thousands of frags.)

Ref. : A.J. Parker, 'Lusitanian amphoras', 38; G. Cardoso, Conimbriga, 17, 1978, 68-69 (no. 11); P.D. Alvarez, Anforas romanas en los caminos del mar, Vigo, 1981, 61-65.

Setúbal (52 from Praça do Bocage; countless frags. from elsewhere)

Ref. : C. Tavares da Silva & J. Soares, Setúbal Arq., 6-7, 1980-81, 262, 265, 267, 268; A. Coelho-Soares & C. Tavares da Silva, Setúbal Arq., 4, 1978, 177-179; A.J. Parker, 'Lusitanian amphoras', 38.

Herdade da Barrosinha, Alcácer do Sal (many frags.) KILN

Ref. : J.-G. Gorges, Les villas hispano-romaines, Paris, 1979, sv. PS 45, p. 484.

SALACIA (Alcácer do Sal) (7 + many frags.)

Ref. : C. Tavares da Silva et al., Setúbal Arq., 6-7, 1980-81, 197.

Cape Sardão (4)

Ref. : G. Cardoso, Conimbriga, 17, 1978, 68 (nos. 12 & 13).

Island of Pessegueiro (15 km SSE of Sines) (many frags.)

Ref. : C. Tavares da Silva, J. Soares, L. Ferrer Dias, Setúbal Arq., 6-7, 1980-81, 224, 244; C. T. da Silva et al., Arq. de Beja, ii, 1, 1984, 41-3.

Aljustrel (2)

Ref. : Conimbriga, 5, 1969, 103 & Pl. XXXVIII, 13 & 14.

EBORA (2)

Ref. : A.J. Parker, art.cit., 38.

Torre de Palma, Monforte (Roman villa) (frags.)

Ref. : A.M. Dias Diogo et al., <u>Série Arq.</u>, Museu do Mar, Cascais, I.4, 1985, 11.

Vila Viçosa (1)

Ref. : A.J. Parker, art.cit., 38.

Quinta do Ribeiro da Nata, Fonte da Moura, Belver, Gavião (Portalegre) (1 handle in C3/4 level)

Ref. : <u>Informação arq.</u>, 5, 1982-83, 95.

Lisbon (Museu Nacional de Arqueologia e Etnologia) (several, provenance uncertain).

Ref. : A.J. Parker, art.cit. 38.

Casilhas (frags.)

Ref. : L. Barros & C. Amaro, <u>Almadan</u>, 4-5, 1984-85, 33-34.

Alcaçova de Santarém (1)

Ref. : A.M. Dias Diogo, <u>Conimbriga</u>, 23, 1984, 126 & 128 (no. 35) & Est. IV.35.

CONIMBRIGA (30)

Ref. : <u>Fouilles de Conimbriga</u>, VI, 80.

2. <u>BAETICA</u>

ITALICA (1)

Ref. : M. Beltran Lloris, <u>Las anforas</u>, 460, fig.

184,8.

GADES (1)

Ref. : A.J. Parker, art.cit. 38 & fig. 13.

3. GALICIA

Bueu, near Pontevedra (1 ?from shipwreck)

Ref. : P.D. Alvarez, Anforas romanas en los caminos del mar, Vigo, 1981, 57-59.

Isorna (Pontecesures, at the mouth of the Ulla) (1)

Ref. : VI Congr. Int. de Arq. Subacuatica Cartagena 1982, Madrid, 1985, 58, fig. 10.

Lugo (Praza de Sto. Domingo) (1)

Variant of S. Bartolomeu de Castro Marim type.

Ref. : F.S. Lopez Gomez, 'A problematica da arq. subacuatica en Galicia', II Seminario Arq. del NO, 1980, Madrid, 1983, 273-90, esp. 285, 287.

Tegra (1)

Ditto

Ref. : ibid.

La Coruña (1)

Ditto

Ref. : ibid.

Castro de Briteiros (1)

Ditto

Ref. : ibid.

4. OTHER PARTS OF HISPANIA TARRACONENSIS

BAETULO (1)

Ref. : M. Comas i Sola, Baetulo. Les amfores, Badalona, 1985, 80-81, 113 (fig. 43).

Freu, near San Feliu de Guixols (1)

Ref. : L. Esteva & R. Pascual, Anales del Instituto de Estudios Gerundenses, 1967, 281-292, fig. 7.

Riells-la Clota, near Ampurias (1 + frags.)

Ref. : F.J. Nieto Prieto & J.M. Nolla, Cypsela, 5, 1985, 150 & fig. IX.2

5. ELSEWHERE

Istres (1)

Ref. : C. Panella, Recherches sur les amphores romaines, Rome, 1972, fig. 26; Ostia III, 520.

Rome (1)

Ref. : Ostia III, 627, no. 14.

OSTIA (Baths of the Swimmer) (Numerous fragments in Antonine and early Severan levels)

Ref. : C. Panella, ibid., figs. 22-23; A. Carandini & C. Panella in A. King & M. Henig (ed.), The Roman in the West in the late Second and Third Centuries, Oxford, 1981, II, 492, 496; Ostia III, 520 & figs. 529, 207, 644, 387, 388, 391, 645, 529.

HERCULANEUM (1)

Ref. : Pompeji, Leben und Kunst in den Vesuvstädten, 1973, 91, no. 4.

6. SHIPWRECKS

Anse de la Fontaine, Javea

Ref. : P. Laugier & F. Carvazé, Cah.d'arch.subaq., 5, 1976, 103 & fig. 11.72 (p. 102).

Conejera Island, San Antonio (Ibiza) (14 + many frags.)

Ref. : B. Vilar Sancho, NAH, 6, 1962, 177-188; B.V. Sancho & J.M. Maña de Angulo, NAH, 7, 1963, 188-194.

Saint Gervais (Fos-sur-Mer) (1 + frags.)

Ref. : B. Liou & R. Marichal, Archaeonautica, 2, 1978, 141 (no. 32 & fig. 16).

Cap Bénat I (25 + frags. + lids)

Ref. : R. Calmes, Cah. d'arch. subaq., 2, 1973, 137-145.

Sud Lavezzi 3, Corsica (1)

Ref. : B. Liou, _Gallia_, 40, 1982, 450.

B. LUSITANIAN GARUM II (= ALMAGRO 50)

1. LUSITANIA

Lagos (many frags.in Museu Regional)

Ref. : A.J. Parker, art.cit., 39 & figs. 23 & 24.

Villa at Mexilhoeira Grande, Abicada (PS 38) (1)

Ref. : J.-G. Gorges, Villas, 481.

BALSA (1 + many frags., some marked)

Ref. : AP, iii, 7-9, 1974-1977, 249-250 & figs. 10,11,13,14,15; CIL II 6254,2; CIL II 6254,33.

Quinta da Alegria (Setúbal) (14 + frags.) KILN

Ref. : A. Coelho-Soares & C. Tavares da Silva, Setúbal Arq., 5, 1979, 205-221.

Tróia (1 + numerous frags.)

Ref. : M. Beltran Lloris, Las anforas, 540; G. Cardoso, Conimbriga, 17, 1978, 71 (no. 22).

Setúbal (4 + many frags.)

Ref. : A. Coelho Soares & C. Tavares da Silva, Setúbal Arq., 4, 1978, 180-181.

SALACIA (Alcácer do Sal) (1)

Ref. : C. Tavares da Silva et al., Setúbal Arq., 6-7, 1980-81, 268.

MIROBRIGA CELTICA (1)

Ref. : A.J. Parker, art.cit., 39 & fig. 22.

Cape Sardão (1)

Ref. : G. Cardoso, Conimbriga, 17, 1978, 71.

Island of Pessegueiro (2 + fragments)

Ref. : C. Tavares da Silva, J. Soares & L.F. Dias, Setúbal Arq., 6-7, 1980-81, 234, 244; C. Tavares da Silva, Almadan, 2, 1983-84, 21.

Vila Viçosa, Museum (1)

Ref. : F. Mayet, MEFRA, 90, 1978, 382.

CONIMBRIGA (1)

Ref. : Fouilles de Conimbriga, VI, 88 (no. 51)

EMERITA (1)

Ref. : R. Lequément, RAN, 9, 1976, 185 (no. 2); F. Mayet, MEFRA, 90, 1978, 382.

2. BAETICA

ONUBA (= Huelva) (1)

Ref. : M. del Amo, Huelva Arq., 2, 1976, 94-95, fig. 41,4 & Pl. 13.

ITALICA (2 marked)

Ref. : F. Mayet, MEFRA, 90, 1978, 382; C. Lamour & F. Mayet, Études sur Pézenas et l'Hérault, 11, 1980, 7, note 21.

3. HISPANIA TARRACONENSIS

CARTHAGO NOVA (1)

Ref. : M. Beltran Lloris, _Las anforas_, 543; F. Mayet, _MEFRA_, 90, 1978, 382.

ELCHE (several frags.)

Ref. : S. Keay, _LRA_, 644.

VALENTIA (1)

Ref. : F. Mayet, _MEFRA_, 90, 1978, 381.

TARRACO : Cathedral Cloister; Palaeochristian Cemetery; Torre de Audiencia (several)

Ref. : C.B. Rüger, _MDAI(M)_, 9, 1968, 257 & fig. 15.4.; M. Beltran Lloris, _Las anforas_, 540 & fig. 220.2; S. Keay, _LRA_, 149-50.

BARCINO (= Barcelona) (3, 1 marked)

Ref. : S. Keay, _LRA_, 149-50, 646.

La Salut, Sabadell (2) villa site

Ref. : S. Keay, _LRA_, 149-50, 644, 646.

Riudarenes, near Constanti

Ref. : S. Keay, _LRA_, 149.

EMPORIAE (5 + frags., some marked; new frs. from town & territory)

Ref. : M. Almagro, _AEA_, 24, 1951, 111 & fig. 9.19; M. Beltran Lloris, _Las anforas_, 540 & fig. 220.1,3,4 (p. 541); S. Keay, _LRA_, 149-50, 169, 646.

POLLENTIA (Majorca)

Ref. : Ostia III, 605.

4. AFRICA

BANASA (1 marked)

Ref. : F. Mayet, MEFRA, 90, 1978, 382-383 & Pl. XIII,2.

Rabat, museum (1 marked)

Ref. : F. Mayet, ibid., 382 & Pl. XIII,1.

Sfax, Tunisia, museum (frags.)

Ref. : Ostia III, 606.

5. GAUL

MASSILIA (Roman docks) (frags.)

Ref. : Ostia III, 606.

NARBO ? (2 marked)

Ref. : C. Lamour & F. Mayet, Études sur Pézenas et l'Hérault, 11, 1980, 9 (notes 95, 97).

6. ITALY

OSTIA (Baths of Swimmer 3 + numerous frags. in levels of late Severan period, later C3 & C4; frags. also in Horrea Epagathiana excavats.)

Ref. : Ostia III, 605, 688, 691, 694, 695; Ostia IV, 121 (figs. 253 & 254), 140-142; CIL XV 3452,2; F. Mayet, MEFRA, 90, 1978, 382; MAAR, 15, 1938,

18.

Rome : Monte Testaccio (1 marked + frags.)

Ref. : CIL XV 3448; Ostia III, 606.

Luni (frags., one marked, in later levels)

Ref. : Scavi di Luni, Rome, 1974, 451 (Tav. 79,9) 455 & (Tav. 111,10); S.L. Siena, 'Appunti su alcuni tipi di anfore lunensi' in Méthodes classiques et méthodes formelles dans l'étude des amphores, 215 & fig. 34.

COSA (frags. in late levels)

Ref. : Ostia IV, 140.

ALBA FUCENS (1 marked)

Ref. : NSA, viii, 6, 1952, 236; F. Mayet, MEFRA, 90, 1978, 382.

Civitavecchia, museum (frags.)

Ref. : Ostia III, 606..

7. EASTERN EMPIRE

SEBASTE (Palestine) (1)

Ref. : A.J. Parker, art.cit., 39.

SAMARIA (1 marked)

Ref. : J.M. & J.W. Crowfoot & K.H. Kenyon, The Objects from Samaria, London, 1957, 388 (no. 4).

Bourgas, museum (Bulgaria) (1)

Ref. : A.J. Parker, art.cit., 39.

8. SHIPWRECKS

Anse Gerbal, Port-Vendres (several)

Ref. : Gallia, 22, 1964, 475 & fig. 2; Y. Chevalier & C. Santamaria, Hommage à F. Benoit, V, 9-10, 32; S. Tortorella, MEFRA, 93, 1981, 373; Ostia III, 628 (fig. 18).

Planier 7 (Marseilles) (several)

Ref. : F. Benoit, Gallia, 20, 1962, 161 & fig. 26.

Pommègues (several)

Ref. : J.-M. Gassend, Archaeonautica, 2, 1978, 101-107.

Anse St. Roche, Antibes (several)

Ref. : F. Benoit, RStLig, 18, 1952, 237-307.

Randello, Camarina (1)

Ref. : S. Keay, LRA, 170, 646.

Cape Ognina (Syracuse) (several)

Ref. : G. Kapitan, Nautical Archaeology, 3.1, 1974, 150; S. Tortorella, MEFRA, 93, 1981, 376.

Marzameni F (Sicily) (several)

Ref. : IJNA 5.4, 1976, 347; Ostia IV, 277-278.

Sud-Lavezzi I (Corsica) (83)

Ref. : IJNA 5.4, 1976, 347; Ostia IV, 277.

C. <u>LUSITANIAN GARUM III</u> (= <u>ALMAGRO 51A-B</u>)

1. <u>LUSITANIA</u>

Quinta da Alegria (Setúbal) (frags.) KILN

Ref. : G. Cardoso, <u>Conimbriga</u>, 17, 1978, 64-5 & Pl. XII; A. Coelho-Soares & C.Tavares da Silva, <u>Setúbal Arq.</u>, 5, 1979, 205-221.

Herdade do Pinheiro (Setúbal) (frags.) KILN

Ref. : G. Cardoso, <u>Conimbriga</u>, 17, 1978, 65 & Pl. XII.

Tróia (many)

Ref. : G. Cardoso, art. cit., 71, 77 (no. 23).

Setúbal (?CAETOBRIGA) (several)

Ref. : A. Coelho-Soares & C. Tavares da Silva, <u>Setúbal Arq.</u>, 4, 1978, 171-201.

SALACIA (several)

Ref. : C. Tavares da Silva et al., <u>Setúbal Arq.</u>, 6-7, 1980-81, 200-201.

Island of Pessegueiro (off Sines) (many)

Ref. : C. Tavares da Silva, <u>Almadan</u>, 2, 1983-84, 21.

2. BAETICA

Cerro del Trigo, Huelva (1)

Ref. : G. Bonsor, MemJSEA, 1928, 17, fig. 8.2; S.J. Keay, LRA, 645, no. 5.

Cartama, Malaga (Roman villa) (3)

Ref. : E. Serrano Ramos & A. De Luque, NAH Arq., 8, 1980, 301, 309.

Colmenar, Malaga (1)

Ref. : S.J. Keay, LRA, 645, no. 12.

3. HISPANIA TARRACONENSIS

Isla de Fraile, Aquilas, Cartagena (?from cetaria)

Ref. : J. Mas, El puerto de Cartagena, 1975, 127, fig. 132.

ELCHE (fragments)

Ref. : S.J. Keay, LRA, 645, no. 7.

Alicante

Ref. : M. Beltran Lloris, Las anforas, 544, fig. 222.5.

TARRACO (32)

Ref. : S.J. Keay, LRA, 157, 645, no. 3.

BARCINO (3)

Ref. : S.J. Keay, LRA, 157, 645, no. 4.

EMPORIAE (4)

Ref. : S.J. Keay, LRA, 157, 645, no. 1.

Roses (11)

Ref. : S.J. Keay, LRA, 157, 645, no. 2.

4. AFRICA

Carthage

Ref. : S.J. Keay, LRA, 168.

5. GAUL

Narbonne : necropolis of St. Paul (1)

Ref. : S.J. Keay, LRA, 159, 645, no. 17.

Saint Seurin, Bordeaux (1)

Ref. : B. Wattier, L' information d'histoire et de l'art, 18, 1973, 113-8; S.J. Keay, LRA, 645, no. 21.

6. ITALY

Rome : Trajan's Market (1)

Ref. : S.J. Keay, LRA, 645, no. 14.

Rome : Schola Praeconum (2)

Ref. : D. Whitehouse et al., PBSR, 50, 1982, 78, fig. 12.166; p. 79, fig. 13.189.

Volterra (1)

Ref. : S.J. Keay, LRA, 645, no. 16.

7. SHIPWRECKS

Favaritz, Cales Coves (Menorca)

Ref. : M. Fernandez-Miranda & M. Belen, _El Fondeadero de Cales Coves_ (= EAE, 101), Madrid, 1979.

Sud-Lavezzi 1 (194)

Ref. : B. Liou, _Gallia_, 40, 1982, 437-44.

Catalans, Marseilles (2)

Ref. : B. Liou, _Gallia_, 31, 1973, 585-86.

Dramont F, Saint Raphael (1)

Ref. : B. Liou, _Gallia_, 33, 1975, 600-01 & fig. 38.

D. LUSITANIAN GARUM IV (= ALMAGRO 51C)

1. LUSITANIA

Quinta da Alegria (Setúbal) (80 + numerous frags.)
KILN

Ref. : A. Coelho-Soares & C. Tavares da Silva, Setúbal Arq., 5, 1979, 205-221; G. Cardoso, Conimbriga, 17, 1978, 71-2 & Pl. XII.

Herdade do Pinheiro, Marateca (Setubal)
(numerous frags.) KILN

Ref. : G. Cardoso, art.cit., 71-72 & Pl. XII; A.J. Parker, art.cit., 36 & fig. 6.

Tróia (numerous frags. on beach + in Museo do Mar, Cascais)

Ref. : M. Beltran Lloris, Las anforas, 630, note 1686; G. Cardoso, art.cit., 71-72, 77 (no. 24).

Setubal (10 + numerous frags.)

Ref. : A. Coelho-Soares & C. Tavares da Silva, Setúbal Arq., 4, 1978, 181-182; C. Tavares da Silva & A. Coelho-Soares, Setúbal Arq., 6-7, 1980-81, 268.

SALACIA (5)

Ref. : C. Tavares da Silva et al., Setúbal Arq., 6-7, 1980-81, 200-201.

Cape Sardao (1)

Ref. : G. Cardoso, art.cit. 71-2, 78 (no. 25).

MIROBRIGA CELTICA (1)

Ref. : A.J. Parker, art.cit., 36 & fig. 3.

Mar de Sardau (between Cape Espichel & Sao Vicente) (2)

Ref. : P.D. Alvarez, Anforas romanas en los caminos del mar, Vigo, 1981, 69-72.

Island of Pessegueiro (off Sines) (numerous)

Ref. : C. Tavares da Silva, J. Soares & L.F. Dias, Setúbal Arq., 6-7, 1980-81, 244; Almadan, 2, 1983-84, 21.

Vila Viçosa, museum (1)

Ref. : A.J. Parker, art.cit., 36 & fig. 4.

Beja, museum (2)

Ref. : A.J. Parker, art.cit., 36 & fig. 5 a-b.

OSSONOBA (Museum at Faro) (1)

Ref. : A.J. Parker, art. cit., 42 & fig. 7.

Lagos, museum (1 marked)

Ref. : F. Mayet, MEFRA, 90, 1978, 362, note 44.

Vale do Boto, S. Bartolomeu de Castro Marim (frags.) KILN

Ref. : V. Gonçalves, H. Catarino & A.M. Arruda, 'O sitio romano-arabe do Vale do Boto', Clio, 2, 1980, 71-79, esp. 72-75; one ? from here in museum at Faro : A.J. Parker, art.cit., 37 & fig. 7.

"Algarve" (1, in museum at Figueira da Foz)

Ref. : ARA, II, fig. 245; Ostia IV, 144.

EMERITA (17)

Ref. : R. Lequément, RAN, 9, 1976, 187; M. Beltran Lloris, Las anforas, fig. 221.

Lisbon, Museu Nacional de Arq. e Etnol. (3)

Ref. : Ostia IV, 143.

Alcaçova de Santarem (1)

Ref. : A.M. Dias Diogo, Conimbriga, 23, 1984, 127 & 129, no. 37 & Est. IV.37.

CONIMBRIGA (5)

Ref. : Fouilles de Conimbriga, VI, 80, 87-88 (nos. 48-49).

AEMINIUM (2)

Ref. : J. de Alarcão, Portugal romano, photo 20; Ostia IV, 143.

2. BAETICA

ONUBA (= Huelva) (frags.)

Ref. : A. del Amo, 'Dos factorias romanas de salazón en Huelva', Huelva Arq., 2, 1976, 23-80, esp. figs. 28 & 34.

El Tejarillo, Lora del Rio, Seville

Ref. : J. Remesal Rodriguez, Saalburg-Jahrbuch, 38, 1982, 30-71, esp. 69.

Seville (museum) (1)

Ref. : S. Keay, LRA, 713, fig. 196.

Calle Marmoles, Malaga

Ref. : S. Keay, LRA, 646.

Cartama, Malaga (Roman villa)

Ref. : S. Keay, LRA, 647.

3. HISPANIA TARRACONENSIS

Almeria, Museo Arq. (1)

Ref. : M. Beltran Lloris, 542 & fig. 221,2; Ostia IV, 143; F. Mayet, MEFRA, 90, 1978, 382.

Uscull Grande (CARTHAGO NOVA) (1)

Ref. : J. Mas, El puerto de Cartagena, Madrid, 1975, 111.

S. Pedro del Pinatar (CARTHAGO NOVA) (1)

Ref. : J.J. de Jauregui, AEA, 21, 1948, 44 & fig. 4.

Elche, Alicante (1)

Ref. : R. Folques, Saitabi, 16, 1975, 31, Pl. V.

PORTUS ILLICITANUS (= Santa Pola) (several)

Ref. : M.J. Sanchez Fernandez & M.T. Lobregad Collado, Lucentum, 3, 1984, 135-51, esp. 146-8 & fig. 9.

VALENTIA (2 + specimens from necropoleis of Les Foies & La Boatella)

Ref. : M. Santandreu Gimeno, Archivo de Preh. Levantina, 11, 1966, 201 & Pl. IVb; Ostia IV, 143;

M. J. Sanchez Fernandez & M.T. Lobregad Collado, Lucentum, 3, 1984, 147.

TARRACO (2)

Ref. : J. Serra Villaro, MemJSEA, 104, 1929, pl. XXXIX, 12; R. Pascual Guasch, Boletin Arq. de Tarragona, 64-65, 1964-1965, 22.

BARCINO (frags.)

Ref. : S. Keay, LRA, 646, no. 2.

Mataro (Municipal museum) (1)

Ref. : S. Keay, LRA, 647, no. 17.

EMPORIAE (frags.)

Ref. : M. Beltran Lloris, Las anforas, 542 & fig. 221,1; M. Almagro Basch, Necropolis de Ampurias, Barcelona, 1953, II, 298 & fig. 269, 302 & fig. 278, 312 & fig. 289.

Can Bosc de Basea

Ref. : S. Keay, LRA, 172, 646, no. 5.

Vilauba

Ref. : ibid., no. 3.

La Guanta, nr. Ampurias

Ref. : ibid., no. 6.

Can Samarruga

Ref. : ibid., no. 7.

Martim, Barcelos (1)

Ref. : Revista da Fac. de Letras de Porto, 3,

1972, 322 & fig. 5.

POLLENTIA (Majorca) (1)

Ref. : M. Vegas, Bonner Jahrb., 165, 1965, 108-40, Pl. 11.7.

4. AFRICA

PORTUS CAESAREA (3)

Ref. : M. Bouchenaki, Fouilles de la nécropole occidentale de Tipasca, Algiers, 1975, 159-60, fig. 103, 2-3 & fig. 105, 13; Ostia IV, 144, 278 & fig. 549.

SABRATHA (1)

Ref. : Ostia IV, 144.

Arrecife, Canaries

Ref. : J. S. Rafols, XI Congreso Nacional de Arqueologia, 1968, Zaragoza, 1970, 428, fig. 2.

5. GAUL

Maureilhan (Herault) (1 marked)

Ref. : C. Lamour & F. Mayet, Études sur Pézenas et l'Hérault, 11, 1980, 9, no. 96.

Beziers (1)

Ref. : Ostia IV, 143.

Saint Raphael (1, ? from sea)

Ref. : J. Joncheray, Nouvelle classification des amphores lors de fouilles sous-marines, 1976, Pl. X. 106.

6. GERMANY

Trier (2 + frags.)

Ref. : S. Loeschcke, Der Tempelbezirk in Altbachtale zu Trier, Berlin, 1942, II, 148, Pl. XXX, no. 36; L. Hussong & H. Cuppers, Trierer Kaiserthermen. Die spätrömische und fruhmittelalterliche Keramik, Mainz, 1972, 23, Pl. VI, no. 54.

7. SICILY

Canale de Sicilia, Palermo (1, from sea)

Ref. : G. Purpura, Sicilia Arch., 35, 1977, 63, fig. 20, Tav. 111.

8. ITALY

Luni

Ref. : S. Keay, LRA, 647, no. 31.

Viareggio (1 from sea)

Ref. : M. Zechini, 'Anfore greche e romane nel mare di Viareggio', La Provincia de Lucca, 11, 1971, 91, fig. 4.

SEMIFONTE (Tuscany)

Ref. : S. Keay, LRA, 647, no. 37.

Civitavecchia (1)

Ref. : P.A. Gianfrotta, 'Archeologia sott'acqua : rinvenimenti sottomarini in Etruria meridionale', Boll. d'Arte, vi, 10, 1981, 80.

OSTIA (Baths of the Swimmer) (Numerous frags. in C3 & C4 A.D. levels)

Ref. : Ostia III, 606; Ostia IV, 142-145.

? Rome (1) (example preserved in Antiquarium Giovampaoliano, Rome, inv. no. 208464)

Ref. : Ostia IV, 144 & fig. 550.

Naples : Via Carminiello ai Mannesi

Ref. : S. Keay, LRA, 647, no. 41.

SYBARIS (frags.)

Ref. : Sybari (NSA 1970, suppl. 3), 498 (fig. 553), 528 (fig. 583, no. 21652), 541 (no. 784).

8. SHIPWRECKS

Les Escolletes, Cabo de Palos, Cartagena

Ref. : J. Mas, VI Congr. Int. de Arq. Subacuatica, 1982, Madrid, 1985, 164, 166 (fig. 7).

Cap Blanc (Majorca) (2)

Ref. : Ostia IV, 278.

Anse Gerbal (Port-Vendres) (several)

Ref. : RStLig, 37, 1971, 11 (fig. 2); S. Tortorella, MEFRA, 93, 1981, 373.

Wreck C, Agde (? is this one wreck or collection of several ?)

Ref. : H. Gallet de Santerre, Archéologie sous-marine, III, 208 (fig. 15); Ostia IV, 144.

Chrétienne D (Marseilles)

Ref. : R. Lequément, RAN, 9, 1976, 187 & fig. 9 c.d.e; Y. Chevalier & C. Santamaria, Hommage à F. Benoit, V, 10-11.

Planier 7 (1)

Ref. : Gallia, 20, 1962, 161 (fig. 27).

Pampelone (Var) (several)

Ref. : R. Lequément, RAN, 9, 1976, 177-188.

Sud Lavezzi I (Corsica) (6)

Ref. : Ostia IV, 277.

Marzamemi F (Sicily) (several)

Ref. : Ostia IV, 277-278; A.J. Parker, IJNA, 10.4, 1981, 328, fig. 19, 12, 3, 4.

Femina Morta (Ragusa, Sicily) (several)

Ref. : S. Tortorella, MEFRA, 93, 1981, 374; A.J. Parker, Kokalos, 22-23, 1976-77, 627, Tav. CXXXV, 10.

B I B L I O G R A P H I E S

I. GENERAL WORKS

ABBOTT, F.F. & JOHNSON, A.C., Municipal Administration in the Roman Empire, Princeton, 1926.

ANDRÉ, J., L'alimentation et la cuisine à Rome (2nd ed.), Paris, 1981.

D'ARMS, J.H., Commerce and Social Standing in Ancient Rome, Cambridge (Mass.), 1981.

D'ARMS, J.H., & KOPFF, E.C. (ed.), The Seaborne Commerce of Ancient Rome : studies in archaeology and history (= MAAR, 36), Rome, 1980.

BLAZQUEZ, J.M. (ed.), Producción y comercio del aceite en la antiguedad (I Congreso Internacional, Madrid, 1980), Madrid, 1981.

BLAZQUEZ, J.M. & REMESAL RODRIGUEZ, J. (ed.), Producción y comercio del aceite en la antiguedad II (II Congreso Internacional, Seville, 1982), Madrid, 1983.

BOUKENAKI, M., Fouilles de la nécropole occidentale de Tipasca, Algiers, 1975.

BOWMAN, A.K. & THOMAS, J.D., Vindolanda : the Latin Writing Tablets, (= Britannia Monograph, 4), London, 1983.

BRAUDEL, F., The Mediterranean and the Mediterranean World in the Age of Philip II (tr. S. Reynolds), London, 1972.

------ The Structures of Everyday Life : the Limits of the Possible (= Civilization and Capitalism, 15th - 18th Century, I), (tr. S. Reynolds), London, 1981.

BRAUNERT, H., . Die Binnenwänderung : Studien zur Sozialgeschichte Ägyptens in der Ptolemäer- und

Kaiserzeit (= Bonner Historische Forschungen, 26), Bonn, 1964.

BRUNT, P.A., 'The revenues of Rome', JRS, 71, 1981, 161-172.

BURNHAM, B.C. & JOHNSON, H.B. (ed.), Invasion and Response : the case of Roman Britain (BAR Brit.Ser., 73), Oxford, 1979.

CALLENDER, M.H., Roman Amphorae, including an index of stamps, London, 1965.

CHARLESWORTH, M.P., Trade Routes and Commerce of the Roman Empire, Hildesheim, 1961.

CHEVALIER, R., Land and Society in Colonial Mexico (tr. A. Eustis), Berkeley, 1963.

COCKLE, H., 'Pottery manufacture in Roman Egypt : a new papyrus', JRS, 71, 1981, 87-97.

CORBIER, M., 'Fiscalité et dépenses locales' in P. Leveau (ed.), L'origine des richesses dépensées en la ville antique (Actes du colloque à Aix-en-Provence, mai 1984), Aix-en-Provence, 1985, 219-232.

CRAWFORD, M.H., Roman Republican Coin Hoards, London, 1969.

---- Roman Republican Coinage (2 vols.), Cambridge, 1974.

---- La moneta in Grecia e a Roma (= Universale Laterza, 615), Rome/Bari, 1982.

---- Coinage and Money under the Roman Republic : Italy and the Mediterranean Economy, London, 1985.

---- 'Money and exchange in the Roman world', JRS,

60, 1970, 40-48.

CROOK, J.A., Law and Life of Rome, London, 1967.

CROWFOOT, J.M. & J.W. & KENYON, K.H., The Objects from Samaria, London, 1957.

DAVIES, R.W., 'The Roman military diet', Britannia, 2, 1971, 122-142.

DRINKWATER, J.F., Roman Gaul, the three provinces 58 B.C. - A.D. 260, Beckenham (Kent), 1983.

DUFF, A.M., Freedmen in the Early Roman Empire, Oxford, 1928.

DUNCAN-JONES, R.P., The Economy of the Roman Empire : Quantitative Studies, Cambridge, 1974.

---- 'The wealth of Gaul', Chiron, 11, 1981, 217-220.

FAVORY, F., 'Le monde des potiers gallo-romains', Les Dossiers d'Archéologie, Sept.-Oct. 1974, 90-102.

FINLEY, M.I., The Ancient Economy, London, 1973.

---- 'Technological innovation and economic progress in the ancient world', EHR, ii, 18, 1965, 29-45.

FORBES, R.J., Studies in Ancient Technology (9 vols.), Leiden, 1955-1965.

FREDERIKSEN, M.W., 'Theory, evidence and the Ancient Economy', JRS, 65, 1975, 164-171.

FRERE, S., Britannia : a history of Roman Britain, London, (2nd ed.), 1975.

FULFORD, M.G., 'The location of Romano-British kilns : institutional trade and the market' in J. Dore & K. Greene (ed.), Roman Pottery Studies in Britain and Beyond (BAR Suppl. Ser., 30), Oxford, 1977, 301-316.

GARNSEY, P.D.A., 'Descendants of freedmen in local politics : some criteria' in B. Levick (ed.), The Ancient Historian and his Materials : essays in honour of C.E. Stevens on his seventieth birthday, Farnborough (Hants.), 1975, 167-180.

---- 'Independent freedmen and the economy of Roman Italy under the principate', Klio, 63, 1981, 359-371.

---- (ed.), Non Slave Labour in the Greco-Roman World (= Cambridge Philological Society, Suppl. Vol. 6), Cambridge, 1980.

GARNSEY, P. & WHITTAKER, C.R. (ed.), Trade and Famine in Classical Antiquity (= Cambridge Philological Society, Suppl. Vol. 8), Cambridge, 1983.

GARNSEY, P., HOPKINS, K. & WHITTAKER, C.R., (ed.), Trade in the Ancient Economy, London, 1983.

GIARDINA, A. & SCHIAVONE, A. (ed.), Merci, Mercati e Scambi nel Mediterraneo (= Societa romana e produzione schiavistica, II), Rome/Bari, 1981.

GRENIER, A., Manuel d' archéologie gallo-romaine. II.2 : Navigation, occupation du sol, Paris, 1934.

HABICHT, Chr., Die Inscriften des Asklepeions (= Altertumer den Pergamon, VIII.3), Berlin, 1969.

HARMAND, J., L'occident romain, Paris, 1960.

HARRIS, W.V., 'Roman terracotta lamps : the organization of an industry', JRS, 70, 1980, 126-145.

HOFFMAN, G.W., A Geography of Europe, including the Asiatic U.S.S.R. (2nd ed.), London, 1961.

HOPKINS, K., 'Economic growth and towns in classical antiquity' in P. Abrams & E.A. Wrigley (ed.), Towns in Societies : essays in economic history and historical sociology, Cambridge, 1978, 35-77.

---- 'Taxes and trade in the Roman Empire', JRS, 70, 1980, 101-125.

HUSSONG, L. & CUPPERS, H., Trierer Kaiserthermen. Die spätrömische und frühmittelalterliche Keramik, Mainz, 1972.

JARMAN, M.R., BAILEY, G.N. & JARMAN, H.N. (ed.), Early European Agriculture : its foundation and development, Cambridge, 1982.

JONES, A.H.M, The Roman Economy (ed. P.A. Brunt), Oxford, 1974.

KOCKEL, V., 'Archäologische Funde und Forschungen in den Vesuvstadten. II', AA, 1986, 443-569.

LANDELS, H.G., Engineering in the Ancient World, London, 1978.

LIEBENAM, W., Stadteverwaltung im römische Kaiserreiche, Leipzig, 1900.

LOESCHCKE, S., Der Tempelbezirk in Altbachtale zu Trier, Berlin, 1942.

MEIGGS, R., Trees and Timber in the Ancient Mediterranean World, Oxford, 1982.

MIDDLETON, P.S., Trade and Market in Roman Imperial Gaul (unpubl. Ph.D. thesis), Cambridge, 1982.

---- 'Army supply in Roman Gaul' in B.C. Burnham & H.B. Johnson (ed.), Invasion and Response : the case of Roman Britain (B.A.R. Brit. Series, 73), Oxford, 1979, 81-97.

MILLAR, F., The Emperor in the Roman World, London, 1977.

---- The Roman Empire and its Neighbours (2nd ed.), London, 1981.

---- 'Condemnation to hard labour in the Roman Empire, from the Julio-Claudians to Constantine', PBSR, 52, 1984, 124-147.

MOMIGLIANO, A.D., Alien Wisdom : the Limits of Hellenization, London, 1975.

MOREL, J.-P., 'La céramique comme indice du commerce antique (réalités et interprétations)' in P. Garnsey & C.R. Whittaker (ed.), Trade and Famine in Classical Antiquity, Cambridge, 1983, 66-74.

OSBORNE, R.G., Demos : the Discovery of Classical Attika, Cambridge, 1985.

PARTRIDGE, C., Skeleton Green (Britannia Monograph, 2), London, 1981.

PATERSON, J.J., '"Salvation from the sea" : amphorae and trade in the Roman West', JRS, 72, 1982, 146-157.

PEACOCK, D.P.S., Pottery in the Roman World : an ethnoarchaeological approach, London, 1982.

PEREIRA MENAUT, G., 'El numero de esclavos en las provincias romanas del Mediterraneo occidental', Klio, 63, 1981, 373-399.

PUCCI, G., 'La produzione della ceramica aretina. Note sull' "industria" nella prima età imperiale', D.d'A.,

7, 1973, 255-293.

---- 'Pottery and trade in the Roman period' in P. Garnsey, K. Hopkins & C.R. Whittaker (ed.), Trade in the Ancient Economy, London, 1983, 105-117.

RICCIONI, G., La terra sigillata della Gallia meridionale (= Problemi di archeologia gallo-romano, I), Bologna, 1977.

RICKMAN, G.E., The Corn Supply of Ancient Rome, Oxford, 1980.

ROMERO MAGALHÃES, J.A., Para o estudo do Algarve economico durante o seculo XVI, Lisbon, 1970.

SHANAHAN, E.N., Western and Central Europe : a regional geography, London, 1964.

SMITH, J.T., 'Villas as a key to social structure' in M. Todd (ed.), Studies in the Romano-British Villa, Leicester, 1978, 149-185.

---- 'Villa plans and social structure in Britain and Gaul', Caesarodunum, 17, 1982 (= Actes du Colloque sur la villa romana dans les provinces de nord-ouest), 321-331.

TCHERNIA, A., Le vin de l'Italie romaine : essai d'histoire économique d'après les amphores (Bibl. des Écoles françaises d'Athènes et de Rome, 261), Rome, 1986.

TREGGIARI, S., Roman Freedmen during the late Republic, Oxford, 1969.

VERTET, H., 'Pauvres potiers, pauvre misère', Les Dossiers d'Archéologie, Sept. - Oct. 1974, 85-89.

WHITTAKER, C.R., 'Inflation and the economy in the fourth century A.D.' in C.E. King (ed.), Imperial Revenue,

Expenditure and Monetary Policy in the Fourth Century A.D. (= BAR Int.Ser., 76), Oxford, 1980, 1-22.

WILSON, A.J.N., *Emigration from Italy in the Republican Age of Rome*, Manchester, 1966.

WILSON, D.R, WRIGHT, R.P. & HASSALL, M.W.C., 'Roman Britain in 1973', *Britannia*, 5, 1974, 397-480.

II. ROMAN SPAIN AND PORTUGAL

ALARCÃO, J. de, _Portugal romano_ (Historia Mundi, 33), (3rd ed.), Lisbon, 1983.

---- 'Sobre a economia rural do Alentejo na época romana', _Conimbriga_, 15, 1976, 5-44.

---- 'A vila romana de S. Cucufate', _Arqueologia (Porto)_, 3, 1981, 117-121.

ALARCÃO, J. DE & ÉTIENNE, R., _Fouilles de Conimbriga. I. L'Architecture_ (2 vols.), _Paris, 1977._

ALBERTINI, E., _Les divisions administratives de l'Espagne romaine_, Paris, 1923.

ALFOLDY, G., _Fasti Hispanienses : senatorische Reichsbeamte und Offiziere in den spanischen Provinzen des römischen Reiches von Augustus bis Diokletian_, Wiesbaden, 1969.

ALMAGRO BASCH, M., _Las estelas decoradas del suroeste peninsular_ (= Bibliotheca Praehistorica Hispana, VIII), Madrid, 1966.

ALMAGRO GORBEA, M., _El Bronce final y el periodo orientalizante en Extremadura_ (= Bibliotheca Praehistorica Hispana, XIV), Madrid, 1977.

---- 'La necropolis de Medellin', _NAH_, 16, 1971, 161-149.

ALMEIDA, D. F. de, _Egitania : Arqueologia e Historia_, Lisbon, 1956.

ALVAREZ MARTINEZ, J.M., _El puente romano de Mérida_ (Monografias Emeritenses, 1), Badajoz, 1983.

---- 'Consideraciones sobre la Mérida prerromana', REstExtr, 40, 1984, 101-109.

AMORIM GIRÃO, A. do, Atlas de Portugal (2nd ed.), Coimbra, 1958.

ARCE, J., El ultimo siglo de la España romana : 284 - 409, Madrid, 1982.

---- 'La "crisis" del siglo III d.C. en Hispania y las invasiones barbaras', Hisp.Ant., 8, 1978, 257-269.

---- 'Inestibilidad politica en Hispania durante el siglo II d.C.', AEA, 54, 1981, 101-115.

ARNAUD, J.M. & GAMITO, T.J., 'Ceramicas estampilhadas da Idade do Ferro do sul de Portugal. I. Cabeça de Vaiamonte, Monforte', AP, iii, 7-9, 1974-77, 165-202.

ARRUDA, M.M., 'Alcaçova de Santarém. Relatorio dos trabalhos arqueologicos de 1984', Clio/Arq., 1, 1983-84, 217-224.

BARRAL I ALTET, X., La circulation des monnaies suèves et visigotiques : contribution à l'histoire économique du royaume visigot (Beihefte der Francia, 4), Munich, 1976.

BEIRÃO, C. DE MELO et al., 'Deposito votivo da II idade do ferro de Garvão : noticia de primeira campanha de escavações', AP, iv, 3, 1985, 45-136.

BLAZQUEZ, J.M., Estructura económica y social de Hispania durante la anarquia militar y el Bajo Imperio, Madrid, 1964.

---- Tartessos y los origines de la colonización fenicia en Occidente (2nd ed.), Salamanca, 1975.

---- La Romanización II. La sociedad y la economia

en la *Hispania romana* (= Ciclos y temas de la historia de Espana), Madrid, 1975.

---- *Historia económica de la Hispania romana*, Madrid, 1978.

---- 'Panorama general de la presencia fenicia y punica en España' in *Atti del I Congresso Internazionale di Studi Fenici e Punici*, Rome, 1983, II, 311-373.

BOUCHIER, E.S., *Spain under the Roman Empire*, Oxford, 1914.

BRAUN, F., *Die Entwicklung der spanischen Provinzialgrenzen im römische Zeit*, Berlin, 1909.

BROUGHTON, T.R.S., 'Oil producing estates in Roman Baetica', *Revista de la Universidad Complutense*, 18, 1979 (= *Homenaje a Garcia y Bellido, IV*), 153-168.

CABO, A. & VIGIL, M., *Historia de España Alfaguara. I. Condicionamentos geograficos. Edad antigua*, Madrid, 1973.

CABRÉ AGUILO, J., *Excavaciones de Las Cogotas, Cardeñosa (Avila). I. El Castro* (= MemJSEA, 110), Madrid, 1930; *II. La necropolis* (= MemJSEA, 120), Madrid, 1932.

CARO BAROJA, J. *Los pueblos de España*, (rev. ed.), 2 vols., Madrid, 1975.

CASTRO HIPOLITO, M. DE, 'Dos tesouros de moedas romanas encontradas em Portugal', *Conimbriga*, 2-3, 1960-61, 1-166.

COELHO, L. 'Inscrições da necropole proto-historica da herdade do Pego - Ourique', *AP*, iii, 5, 1971, 167-180.

COLLANTES VIDAL, E., 'Reacuñaciones en la moneda iberica' Ampurias, 31-32, 1969-70, 255-257.

CUQ, E., 'Notes d'épigraphie et de papyrologie juridiques', NRDHDFE, 1908, 285-312.

DELGADO ALVES, L.F., 'Aspectos de arqueologia em Myrtilis', Arquivo de Beja, 13, 1956, 41-45.

DUNCAN-JONES, R.P., 'The procurator as civic benefactor', JRS, 64, 1974, 79-85.

EDMONDSON, J.C., Economy and Society of Roman Lusitania under the Principate (unpubl. Ph.D. thesis, Cambridge, 1984.

---- 'Mithras at Pax Iulia - a re-examination', Conimbriga, 23, 1984, 69-86.

ENCARNAÇÃO, J. de, Inscrições romanas do Conventus Pacensis : subsidios para o estudo da romanização, Coimbra, 1984 [= IRCP]

ÉTIENNE, R., Le culte impérial dans la péninsule ibérique d'Auguste à Diocletien, Paris, 1958.

---- 'Sénateurs originaires de la province de Lusitanie' in S. Panciera (ed.), Epigrafia e Ordine Senatorio (Colloquio internazionale, Rome, 1981), (= Tituli, 5), Rome, 1982, 521-529.

FABRE, G., 'Le tissu urbain dans le nord-ouest de la péninsule ibérique', Latomus, 29, 1970, 314-339.

GALSTERER, H., Untersuchungen zum römischen Städtewesen auf der iberischen Halbinsel (= Madrider Forschungen, 8), Berlin, 1971.

GARCIA IGLESIAS, L.A., 'La Béturia : un problema geográfico de la Hispania antigua', AEA, 44, 1971,

86-108.

---- 'El Guadiana y los limites comunes de Bética y Lusitania', Hisp.Ant., 2, 1972, 165-177.

GARCIA Y BELLIDO, A., Fenicios y Cartagineses en Occidente, Madrid, 1942.

---- España y los Españoles hace dos mil anos según la geografia de Estrabón (6th ed.), Madrid, 1978.

---- La España del siglo primero de nuestra era (segun P. Mela y C. Plinio), (3rd ed.), Madrid, 1978.

---- 'Las colonias romanas de la provincia Lusitania', Arqueologia e Historia, 8, 1958, 13-23.

---- 'El recinto mural romano de Evora', Conimbriga, 10, 1971, 85-92.

GIL DA CRUZ SOARES MONTANS, V., 'Notas acerca tres inscrições de Olisipo', Conimbriga, 15, 1976, 151-169.

GONCALVES, V., CATARINO, H., ARRUDA, A.M., 'O sitio romano-arabe do Vale de Boto', Clio, 2, 1980, 71-79.

GORGES, J.-G., Les villas hispano-romaines : inventaire et problématique archéologiques (Publ. du Centre Pierre Paris, 4), Paris, 1979.

---- 'Centuriation et organisation du térritoire : notes préliminaires sur l'exemple de Mérida' in P.A. Fevrier & P. Leveau (ed.), Villes et campagnes dans l'Empire romain, Aix-en-Provence, 1982, 101-110.

GUADAN, A.M. de, La moneda iberica : catálogo de numismática ibérica e ibero-romana, Madrid, 1980.

HURTADO DE SAN ANTONIO, R., Corpus Provincial de

Inscripciones Latinas - Cáceres, Cáceres, 1977 [= CPILC]

JONES, R.F.J., 'The Roman military occupation of north-west Spain', JRS, 66, 1976, 45-66.

KING, P.D., Law and Society in the Visigothic Kingdom, Cambridge, 1972.

LAMBRINO, S., 'Catalogue des inscriptions latines du Musée Leite de Vasconcelos', AP, iii, 1, 1967, 123-217.

LASSERRE, F., Strabon. Geographie II (livres III & IV) (Bude ed.), Paris, 1966.

---- 'Strabon devant l'Empire romain', ANRW, II.30.1, Berlin/New York, 1982, 867-896.

MACKIE, N., Local Administration in Roman Spain A.D. 14 - 212 (BAR Int. Ser., 172), Oxford, 1983.

MAIA, M., 'Fortalezas romanas do sul de Portugal', Zephyrus, 28-29, 1978, 279-285.

---- 'Povos do sul de Portugal nas fontes classicas - Celtici e Turduli', Clio, 2, 1980, 67-70.

---- 'Povos do sul de Portugal nas fontes classicas - os Conios' in Actas do IV Congresso Nacional de Arqueologia (Faro, 1980) (forthcoming)

MOITA, I., 'Problemas de Lisboa romana. A recuperação do teatro de Olisipo' in Arqueologia de las ciudades modernas superpuestas a las antiguas (Zaragoza, 1983), Madrid, 1985, 285-302.

OLIVEIRA MARQUES, A.H. DE, Daily Life in Portugal in the late Middle Ages (tr. S.S. Wyatt), Wisconsin, 1971.

D'ORS, A., Epigrafia juridica de la España romana, Madrid, 1953.

NUNES RIBEIRO, F., 'Pré-historia e a origem de Beja' Arq. de Beja, 17, 1960, 3-113.

PASTOR MUNOZ, M., 'La peninsula iberica en Marciano de Heraclea', Hisp. Ant., 8, 1978, 89-128.

PEREIRA, I., BOST, J.-P. & HIERNARD, J., Fouilles de Conimbriga (ed. J. de Alarcão & R. Étienne). III. Les Monnaies, Paris, 1974.

DA PONTE, S., 'Tomar - antiga Sellium', Arqueologia (Porto), 11, jun. 1985, 112-114.

RADDATZ, K., Die Schatzfunde der iberischen Halbinsel von Ende des dritten bis zur Mitte des ersten Jahrhunderts von Chr. Geb. : Untersuchungen zur hispanischen Toreutik (= Madrider Forschungen, 5), (2 vols.), Berlin, 1969.

REMESAL RODRIGUEZ, J., La annona militaris y la exportación de aceite bético a Germania : con un corpus de sellos en anforas Dressel 20 hallados en Nimega, Colonia, Mainz, Saalburg, Zugmantel y Nida-Heddernheim, Madrid, 1986.

---- 'La economia oleicola betica : nuevas formas de analisis', AEA, 50-51, 1977-78, 87-142.

RICHMOND, I.A., 'The first years of Emerita Augusta', AJ, 87, 1930, 98-116.

ROLDAN HERVAS, J.M., Hispania y el ejercito romano : contribución a la historia social de la España antigua (= Acta Salmanticensia, 76), Salamanca, 1974.

---- 'Fuentes antiguas para el estudio de los Vettones', Zephyrus, 19-20, 1968-1969, 73-106.

LE ROUX, P., L'armée romaine et l'organization des provinces ibériques d'Auguste à l'invasion de 409 (= Publication du Centre Pierre Paris, 8), Paris, 1982.

---- 'L'armée de la péninsule ibérique et la vie économique sous le Haut-Empire romain' in Armées et fiscalité dans le monde antique (Colloque du CNRS), Paris, 1977, 341-372.

ROYAL NAVAL INTELLIGENCE DIVISION, Spain and Portugal (Geographical Handbook Series), 3 vols., London, 1941.

DOS SANTOS ROCHA, A., 'Noticia de algumas estações romanas e arabes do Algarve', AP, 1, 1895, 193-212.

SCHUBART, H., 'Phönizische Niederlassungen an der iberische Sudkuste' in Phönizier im Westen (= Madrider Beitrage, 8), Mainz, 1982, 207-234.

SCHULTEN, A., Iberische Landeskunde : Geographie des antiken Spaniens, Strasbourg, (2 vols.), 1955, 1957.

____ 'Hispania' in RE VIII. 1965-2046.

SILLIÈRES, P., 'Centuriation et voie romaine au sud de Mérida : contribution à la délimination de la Bétique et de la Lusitanie', MCV, 18, 1982, 437-448.

SOARES, J. & TAVARES DA SILVA, C., 'Ocupação do periodo proto-romano do povoado do Pedrão (Setubal)', Actas das II Jornadas Arq., 1972, I, 245-306.

SYME, R., Colonial Elites : Rome, Spain and the Americas, Oxford, 1958.

---- 'The conquest of north-west Spain' in Legio VII Gemina, León, 1970, 83-107 = Roman Papers (ed. E. Badian), II, Oxford, 1979, 825-854.

TAVARES DA SILVA, C., 'Escavações arqueologicas na Ilha do Pessegueiro', Almadan, 2, 1983-84, 20-22.

TAVARES DA SILVA, C., SOARES, J., BEIRÃO, C. DE MELO, FERRER DIAS, L. & COELHO-SOARES, A., 'Escavações arqueologicas no Castelo de Alcácer do Sal (campanha de 1979)', Setúbal Arq., 6-7, 1980-81, 149-218.

TAVARES DA SILVA, C., SOARES, J. & FERRER DIAS, L., 'Trabalhos arqueologicas na Ilha do Pessegueiro (1980)', Setúbal Arq., 6-7, 1980-81, 219-247.

TAVARES DA SILVA, C., SOARES, J., FERRER DIAS, L. & COELHO-SOARES, A., 'Escavações arqueologicas na Ilha do Pessegueiro (Sines) : noticia de 2a. campanha (1981)', Arq. de Beja, ii, 1, 1984, 11-45.

THOMPSON, E.A., The Goths in Spain, Oxford, 1969.

THOUVENOT, R., Essai sur la province romaine de Bétique, Paris, 1940.

TOVAR, A., Iberische Landeskunde II. Die Völker und die Städte des antiken Hispanien. 2. Lusitanien, Baden-Baden, 1976.

TRANOY, A., La Galice romaine : recherches sur le nord-ouest de la péninsule ibérique dans l'antiquité (= Publication du Centre Pierre Paris, 7), Paris, 1981.

ULBERT, G., Cáceres el Viejo : ein spätrepublikanisches Legionslager in Spanisch Extremadura (Madrider Beitrage, 11), Mainz am Rhein, 1984.

VAN NOSTRAND, J.J. 'Roman Spain' in T. Frank (ed.), An Economic Survey of Ancient Rome, III, Baltimore, 1937, 119-224.

VARELA GOMES, R.M., 'Uma estela epigrafada da Herdade dos Franciscos (Garvão)', Conimbriga, 23, 1984, 43-54.

VASCONCELOS, J. LEITE de, 'Da Lusitania a Betica', AP, 5, 1900, 225-249.

VEIGA, E. da, 'Antiguedades monumentães do Algarve', AP, 15, 1910, 209-233.

VEIGA AFFONSO DOS SANTOS, M.L.E. DA, Arqueologia romana do Algarve (2 vols.), Lisbon, 1971, 1972.

VEIGA FERREIRA, O. da, 'Ara votiva de Lousă', Rev. Guim., 62, 1952, 192-195.

VEIGA FERREIRA, S. da, 'Marcas de oleiro em territorio português', AP, iii, 3, 1969, 131-177.

VIEIRA DA SILVA, A., Epigrafia de Olisipo : subsidios para a historia de Lisboa romana, Lisbon, 1944.

VIVES, J., Inscripciones latinas de la España romana (2 vols.), Barcelona, 1971, 1972 [=ILER].

VIVES Y ESCUDERO, A., La moneda hispanica (4 vols.), Madrid, 1924.

WEST, L.C., Roman Imperial Spain : the Objects of Trade, Oxford, 1929.

WICKERT, L., 'Bericht über eine zweite Reise zur Vorbereitung von CIL II Suppl. 2', Sitz. der Preuss. Ak. der Wissen., 1931, 829-844.

WIEGELS, R., 'Zum Territorium des augusteischen Kolonie Emerita', MDAI(M), 17, 1976, 258-284.

WISEMAN, F.J., Roman Spain, London, 1956.

III. MINING

ALARCÃO, J. de & A., 'O espolio da necropole luso-romana de Valdoca (Aljustrel)', Conimbriga, 5, 1966, 7-104.

ALLAN, J.C., Considerations on the Antiquity of Mining in the Iberian Peninsula (= Royal Anthropological Institute, Occasional Paper, 27), London, 1970.

----- 'A mineração em Portugal na antiguedade', Boletim de Minas, 2, 1965, 139-75.

ALMEIDA, D. F. de, 'Mineração romana em Portugal' in La Mineria Hispana e Iberoamericana, León, 1970, I, 195-220.

----- 'Minas de ouro na Gallaecia portuguesa' in Legio VII Gemina, León, 1970, 287-301.

D'ANDRADE R. F., 'A lavra romana das minas de Algares e na herdade do Montinho', Actas do I Congreso Nacional de Arqueologia, Lisboa, 1958, Lisbon, 1970, II, 273-283.

-------- 'Uma lapida romana do Monte do Farrobo - Rio de Moinhos', Conimbriga, 1, 1959, 109-114.

-------- 'Documentos ineditos para a historia das minas de Aljustrel no século XIX', Arq. de Beja, 23-24, 1966-1967, 337-352.

BAKEWELL, P.J., Silver Mining and Society in Colonial Mexico : Zacatecas 1546 - 1700 (= Cambridge Latin American Studies, 15), Cambridge, 1971.

BEIRÃO, C. DE MELO, ALVES DIAS, M.M. & COELHO, L., 'Duas necropoles da Idade do Ferro no Baixo Alentejo : Ourique (noticia preliminar)', AP, iii, 4, 1970, 175-219.

BELCHOIR, C., 'Duas notas sobre lucernas da necropole romana de Valdoca (Aljustrel)', Conimbriga, 9, 1970, 75-79.

BENSON, J., British Coalminers in the Nineteenth Century : a social history, Dublin, 1980.

BERGER, A. 'A labor contract of A.D. 164', CPh, 43, 1948, 231-242.

BLANCO FREIJEIRO, A, 'Antigüedades de Riotinto', Zephyrus, 13, 1962, 31-45.

--- & LUZON NOGUE, J.M., 'Mineros antiguos españoles', AEA, 39, 1966, 73-88.

--- & ---, 'Pre-Roman silver mines at Riotinto', Antiquity, 43, 1969, 124-131.

BLAZQUEZ, J.M., 'Explotaciones mineras en Hispania durante la Republica y el alto Imperio romano : problemas económicos, sociales y tecnicos', AHES, 2, 1969, 9-68.

------- 'Fuentes literarias griegas y romanas referentes a las explotaciones mineras de la Hispania romana' in La Mineria Hispana e Iberoamericana, León, 1970, I, 117-150.

BORLASE, W., Tin Mining in Spain, past and present (= Ancient Science Tracts 429), London, 1897.

BOULAKIA, J.D.C., 'Lead in the Roman world', AJA, 76, 1972, 139-144.

BOON, G. C., 'Aperçu sur la production des métaux non ferreux dans la Bretagne romaine', Apulum, 9, 1971, 453-503.

BRADING, D. & CROSS, H., 'Colonial silver mining : Mexico and Peru', Hispanic American Historical Review, 52, 1972, 545-579.

CANTO, A.M., 'Avances sobre la explotación del marmor en la España romana', AEA, 50-51, 1977-1978, 165-188.

CHECKLAND, S.G., The Mines of Tharsis : Roman, French and British Enterprise in Spain, London, 1967.

CLEERE, H., 'Some operating parameters for Roman ironworks', Bulletin of Instit. of Arch., 13, 1976, 233-246.

COLLS, D., DOMERGUE, Cl., LAUBENHEIMER, F. & LIOU, B., 'Les lingots d'étain de l'épave Port-Vendres II', Gallia, 33, 1975, 61-94.

DAVIES, O., Roman Mines in Europe, Oxford, 1935. [= RME]

----- 'Ancient Roman and medieval mining technique', Trans. Inst. Min. Metal, 43, 1934, 3-54.

DIAZ Y DIAZ, M.C., 'Metales y mineria en la epoca visigotica a traves de Isidoro de Sevilla' in La Mineria Hispana e Iberoamericana, León, 1970, 261-274.

DIRECCAO-GERAL DE MINAS E SERVICIOS GEOLOGICOS, Carta Mineira de Portugal (1:500,000), Lisbon, 1960.

DOMERGUE, Cl., La mine antique d'Aljustrel (Portugal) et les tables de bronze de Vipasca (= Publication du Centre Pierre Paris, 9), Paris, 1983.

------ Les mines de la péninsule ibérique dans l'antiquité romaine (unpubl. Ph.D. thesis, Univ. de

Paris : Sorbonne), 1979 (to be published).

------ Catalogue des mines et des fondéries antiques de la péninsule ibérique (forthcoming, Madrid, 1987).

------- Les lingots de plomb romains estampillés d'Espagne (forthcoming).

------- 'Les Planii et leur activité industrielle en Espagne sous la république', MCV, 1, 1965, 9-27.

------- 'Les lingots de plomb romains du Musée archéologique de Carthagène et du Musée Naval de Madrid', AEA, 39, 1966, 41-72.

------ 'Un temoignage sur l'industrie minière et métallurgique du plomb dans la region d'Azuaga (Badajoz) pendant la guerre de Sertorius', XI Congreso Nacional de Arqueologia, Mérida 1968, Zaragoza, 1970, 608-626.

------ 'Introduction à l'étude des mines d'or dans le nord-ouest de la péninsule ibérique' in Legio VII Gemina, León, 1970, 255-286.

------ 'Rapports entre la zone minière de la Sierra Morena et la plaine agricole de Guadalquivir à l'époque romaine', MCV, 8, 1972, 614-622.

------ 'A propos de Pline, "Naturalis Historia", 33.70-78 et pour illustrer sa description des mines d'or romaines d'Espagne', AEA, 45-47, 1972-1974, 499-528.

------ 'L'eau dans les mines d'or romaines du nord-ouest de l'Espagne' in P. Louis (ed.), L'homme et l'eau en Méditerranee et au Proche Orient. III : L'eau dans les techniques (= Travaux de la Maison de l'Orient, 11), Lyons, 1986, 109-119.

--- & D'ANDRADE, R. F., 'Sondages 1967 et 1969 à Aljustrel (Portugal) : note préliminaire',

Conimbriga, 10, 1971, 99-116.

--- & HERAIL, G., *Mines d'or romaines d'Espagne : le district de la Valduerna (Leon) : étude géomorphologique et archéologique* (2 vols.), Toulouse, 1978.

--- & SILLIÈRES, P., *Minas de oro romanas de la provincia de León I. La Corona de Quintanilla : excavaciones 1971-1973; Las Coronas de Filliel, Boisan, Luyego I & II : exploraciones 1973* (= EAE, 93), Madrid, 1977.

--- & MARTIN, T., *Minas de oro de la provincia de León II* (= EAE, 94), Madrid, 1977.

DUSANIC, S., 'Roman mining in the Danubian provinces', *ANRW*, II.6, Berlin/New York, 1977, 53-94.

EDMONDSON, J.C., 'Mining in the later Roman Empire : continuity or disruption ?' (forthcoming).

EUZENNAT, M., 'Lingots espagnols retrouvés en mer', *Études classiques (Publication de l'université d'Aix-en-Provence)*, 3, 1968-1969, 83-98.

FERNANDEZ NIETO, F. J., '"Aurifer Tagus"', *Zephyrus*, 21-22, 1970-1971, 245-259.

FLACH, D., 'Die Bergwerksordnungen von Vipasca', *Chiron*, 9, 1979, 399-448, esp. 440-443.

GARCIA Y BELLIDO, M.P., *Las monedas de Castulo en escritura indigena : historia numismatica de una ciudad minera*, Madrid, 1982.

HEALY, J.F., *Mining and Metallurgy in the Greek and Roman World*, London, 1978.

------ 'Problems in mineralogy and metallurgy in Pliny the Elder's Natural History' in *Tecnologia* economica e societa nel mondo romano, Como, 1981, 163-201, esp. 184-186.

HOPPER, R.J., 'The Attic silver mines in the fourth century B.C.', ABSA, 48, 1953, 200-254.

------ 'The Laurion mines : a reconsideration', ABSA, 63, 1968, 293-326.

JONES, G.D.B., 'Roman mines at Riotinto', JRS, 70, 1980, 146-165.

JONES, R.F.J. & BIRD, D. G., 'Roman gold-mining in north-west Spain. II', JRS, 62, 1972, 59-74.

LAUBENHEIMER-LEENHARDT, F. & GALLET DE SANTERRE, H., Recherches sur les lingots de cuivre et de plomb d'époque romaine dans les régions de Languedoc - Roussillon et de Provence - Corse (= Rév. arch. de Narbonnaise, Suppl. 3), Paris, 1973.

LEWIS, P. R. & JONES, G.D.B., 'The Dolaucothi gold mines I : the surface evidence', Antiq. Journ., 49, 1969, 244-272.

------- 'Roman gold mining in north-west Spain', JRS, 60, 1970, 169-185.

LUZON NOGUE, J.M., 'Lucernas romanas de Riotinto', AEA, 40, 1967, 138-150.

-------- 'Los sistemas de desagüe en minas romanas del suroeste peninsular', AEA, 41, 1968, 101-120.

--- & RUIZ MATA, D., 'El poblado minero de Riotinto', Habis, 1, 1970, 125-138.

MACLAREN, J.M., Gold : its geological occurrence and geographical distribution, London, 1908.

MAGUEIJO, C., 'A Lex Metallis Dicta 117-138', AP, iii, 4, 1970, 125-163.

MALUQUER DE MOTES, J., 'Excavaciones arqueologicas en el castro de "Las Merchanas" (Lumbrales, Salamanca)', Pyrenae, 4, 1968, 101-128.

MANNING, W.H., 'The native and Roman contribution to the development of metal industries in Britain' in B.C. Burnham & H.B. Johnson (ed.), Invasion and Response (BAR Brit. Ser., 73), Oxford, 1979, 111-121.

DE MESA, J. & GISBERT, T., 'Oruro : origem de una villa minera' in La Mineria Hispana e Iberoamericana, León, 1970, I, 559-590.

MONTEAGUDO, L., 'Metalurgia hispana de la edad del Bronce con especial estudio de Galicia y norte de Portugal', Caesaraugusta, 4, 1954, 55-95.

MROZEK, S., 'Die Goldbergwerke im römischen Dazien', ANRW, II.6, Berlin/New York, 1977, 95-109.

D'ORS, A., 'La condición juridica del suelo en las provincias de Hispania' in I diritti locali nelle province romane con particolare riguardo alle condizioni giuridiche del suolo (= Accad. Nazion. dei Lincei, 194), Rome, 1974, 253-268.

PARREIRA, R., 'Intervenção de emergencia no povoado mineiro romano de Vipasca - Aljustrel, distrito de Beja', Informação arq., 4, 1981, 135-144.

------ 'O salvamento arqueologico de Vipasca : nota preliminar sobre a campanha de escavações de 1981', Arq. de Beja, ii, 1, 1984, 83-106.

PATTERSON, C.C., 'Silver stocks and losses in ancient and medieval times', EHR, ii, 25, 1972, 205-235.

RICHARDSON, J.S., 'The Spanish mines and the development of provincial taxation in the second century B.C.', *JRS*, 66, 1976, 139-152.

RICKARD, T.A., 'The mining of the Romans in Spain', *JRS*, 18, 1928, 129-143.

ROTHERBERG, B. & BLANCO FREIJEIRO, A., *Studies in Ancient Mining and Metallurgy in South-West Spain : explotations and excavations in the province of Huelva* (= Metal in History, 1), London, 1981.

ROUX, P. le, 'Procurator affranchi in Hispania : Saturninus et l'activité minière', *MDAI(M)*, 26, 1985, 218-233.

SAENZ FERNANDEZ, P., 'Metalurgia y comercio punicos en Sierra Morena' in *Homenaje al Prof. Dr. Hernandez Diaz*, I, Seville, 1982, 105-115.

SALKIELD, L.U., 'Ancient slags of the south-west of the Iberian peninsula' in *La Mineria Hispana e Iberoamericana*, Leon, 1970, I, 85-98.

SANCHEZ PALENCIA, F.J., 'La explotación del oro en la Hispania romana : sus inicios y precedentes', *I Congreso Internacional de Mineria Antigua en el Mundo Mediterraneo (Madrid, Oct. 1985)* (forthcoming).

TEIXIERA, C., 'Minas romanas na Serra da Lousã', *Trab. de Antr. e Etnol.*, 10, 1945-1946, 243-247.

VAN NOSTRAND, J.J., 'Two notes on the mines of Roman Spain', *Pacif. Hist. Rev.*, 4, 1935, 281-289.

VASCONCELOS, J. LEITE de, 'Excursão arqueologica a Extremadura Transtagana', *AP*, 19, 1914, 300-323.

VEIGA FERREIRA, O. da & d'ANDRADE, R.F., 'A necropole de Valdoca (Aljustrel)', *Conimbriga*, 5, 1966, 1-6.

VENY, C., 'Diecesiete lingotes de plomo de una nave romana de Ses Salines (Mallorca)', Ampurias, 31-32, 1969-1970, 191-219.

VIANA, A., VEIGA FERREIRA, O. da & d'ANDRADE, R.F., 'Exploração das minas de Aljustrel pelos romanos', Arq. de Beja, 13, 1956, 3-20.

IV. GARUM AND COMMERCE

ABREU NUNES, M.L., 'Tesouro de moedas romanas encontradas em Tróia', AP, iii, 7-9, 1974-1977, 359-364.

ALARCÃO, A., BELOTO, C., ENCARNAÇÃO, J. de, DE ALMEIDA, M.M., 'O mosaico do Oceano de Faro', Anais do Municipio de Faro, 10, 1980, 219-232.

ALEXANDER, J., 'The salt industries of Africa : their significance for European prehistory' in K.W. de Brisay & K.A. Evans (ed.), Salt : the study of an ancient industry (Report on the Salt Weekend held at the University of Essex, 20-22 Sept. 1974), Colchester, 1975, 81-2.

ALMEIDA, D.F. de & CAVALEIRO PAIXÃO, A., Notas sobre a estação arqueologica de Tróia de Setúbal, Setúbal, 1978.

-----, ZBYSZEWSKI, G. & VEIGA FERREIRA, O. da, 'Descoberta de fornos lusitano-romanos na região de Marateca (Setúbal)', AP, iii, 5, 1971, 155-166.

AMARO, C., 'Casa dos Bicos - noticia historico-arqueologico', Arqueologia (Porto), 6, dezembro 1982, 96-111.

BALIL, A., 'Navigazione e rapporti maritimi nella Spagna romana' in Philias Charin. Miscellanea di studi classici in onore di Eugenio Manni, Rome, 1979, I, 171-180.

--- 'Hispania y Ostia', AEA, 33, 1960, 215-217.

BELTRAN LLORIS, M., Las anforas romanas de España (= Monografias arqueologicas, VIII), Zaragoza, 1970.

BENOIT, F., 'Nouvelles épaves de Provence (III)', Gallia,

20, 1962, 147-176.

BERNARDO, H. de BARROS, 'Marinhas ignoradas da Estremadura - as salinas de Peniche', Ethnos, 5, 1966, 121-167.

BESNIER, M., 'Sal' in Daremberg-Saglio (ed.), Dictionnaire des antiquités grecques et romaines, 4.2, 1009-1012.

--- 'Purpura', ibid., 4.1, 769-779.

BORGSTROM, G. & HEIGHWAY, A.J., Atlantic Ocean Fisheries, London, 1961.

BOURGE, J., 'Étude sur les migrations du "scomber thynnus" (thon commun) dans le bassin occidental de la Méditerranée', Révue Tunisienne, 1915, 85-97.

BROUGHTON, T.R.S., 'Some notes on trade and traders in Roman Spain' in J.A.S. Evans (ed.), Polis and Imperium : studies in honour of E.T. Salmon, Toronto, 1974, 11-30.

CABEÇADAS, H., SOARES, J. & GUERRA, A.J. (ed.), Embarcações tradicionais do Sado, Setúbal, 1986.

CARANDINI, A. & PANELLA, C. (ed.), Ostia III (= Studi miscellanei, 21), Rome, 1973.

-------- Ostia IV (= Studia miscellanei, 23), Rome, 1977.

CARANDINI, A. & PANELLA, C. 'The trading connexions of Rome and central Italy in the late second and third centuries : the evidence of the Terme del Nuotatore excavations, Ostia' in A. King & M. Henig (ed.), The Roman West in the Third Century (= BAR Int.Ser., 109), Oxford, 1981, 487-503.

CARDOSO, G., 'Anforas romanas do Museu do Mar (Cascais)', Conimbriga, 17, 1978, 63-78.

CASSON, L., 'Harbour and river boats of ancient Rome', JRS, 55, 1965, 31-39.

CHARLIN, G., GASSEND, J.-M. & LEQUÉMENT, R., 'L'épave antique de la baie de Cavalière (Le Lavandrou, Var)', Archaeonautica, 2, 1978, 9-93.

CHEVALIER, Y. & SANTAMARIA, C., 'L'épave de l'Anse Gerbal à Port-Vendres' in Omaggio a Fernand Benoit, Bordighera, 1971, V, 7-32.

COELHO SOARES, A. & TAVARES DA SILVA, C., 'Anforas romanas da area urbana de Setúbal', Setúbal Arq., 4, 1978, 171-201.

---- & ---- , 'Anforas romanas da Quinta da Alegria (Setúbal)', Setúbal Arq., 5, 1979, 205-221.

COLLS, D., ÉTIENNE, R., LEQUÉMENT, R., LIOU, B. & MAYET, F., L'épave Port-Vendres II et le commerce de la Bétique à l'époque de Claude (= Archaeonautica, 1), Paris, 1977.

DA COSTA, J.P., 'Estudo da fauna malacologica no espolio da gruta natural de Ibne Amar', Actas do II Congreso Nacional de Arqueologia, Coimbra 1970, Coimbra, 1971, II, 599-615.

COULL, J.R., The Fisheries of Europe : an economic geography, London, 1972.

COUTINHO, C, 'Fabrica romana com 2000 anos surge de baixo do asfalto em Cacilhas', O Diario, 17th Dec. 1981, 5.

CURTIS, R.I., Production and Commerce of Fish Sauce in the Western Roman Empire : a social and economic study (unpubl. Ph.D thesis), Maryland, 1978.

--- 'The garum shop of Pompeii (I.12,8)', Cron. Pomp., 5, 1979, 5-23.

--- 'In defense of garum', CJ, 78, 1983, 232-40.

--- 'The salted fish industry of Pompeii', Archaeology, 37.6, Nov./Dec. 1984, 58-9, 74-5.

--- 'Negotiatores allecarii and the herring', Phoenix, 38, 1984, 147-158.

CUTTING, C.L., Fish Saving : a history of fish-processing from ancient to modern times, London, 1955.

DESJACQUES, P. & KOEBERLÉ, P. 'Mogador et les Îles Purpuraires', Hesperis, 42, 1955, 193-202.

DIAS DIOGO, A.M., 'Fornos de anforas do Monte do Bugio. Noticia preliminar', Conimbriga, 19, 1980, 147-150.

------- 'Fornos de anforas do Monte da Enchurrasqueira e do Vale da Cepa - nota preliminar', Conimbriga, 22, 1983, 209-215.

------- O material dos fornos romanos da Enchurrasqueira no Museu do Mar, Cascais (= Série Arq., Museu do Mar, Cascais, I.3), Cascais, 1984.

------- O material dos fornos romanos do Abul no Museu do Mar, Cascais (= Série Arq., Museu do Mar, Cascais, I.4), Cascais, 1985.

DIAZ ALVAREZ, P., Anforas romanas en el eje atlantico galaeco-lusitano, Vigo, 1984.

DIDSBURY, B., 'Cheshire saltworkers' in R. Samuel (ed.), Miners, Quarrymen and Saltworkers, London, 1977, 137-203.

DRESSEL, E., 'Di un grande deposito di anfore rinvenuto nel nuovo quartiere del Castro Pretorio', BCAR, 7, 1879, 36-112; 143-196.

DUMONT, J., 'La peche du thon à Byzance a l'époque hellénistique', REA, 78-79, 1976-77, 96-119.

DUNCAN-JONES, R.P., 'Giant cargo ships', CQ, 27, 1977, 331-332.

ÉTIENNE, R., 'Les amphores du Testaccio au IIIe. siècle', MEFRA, 61, 1949, 151-181.

---- 'Ausone et l'Espagne' in Mélanges d'archéologie, d'épigraphie et d'histoire offerts à Jérome Carcopino, Paris, 1966, 319-332.

---- 'À propos du "garum sociorum"', Latomus, 29, 1970, 297-313.

EVANS, J.D., 'The exploitation of molluscs' in P.J. Ucko & G.W. Dimbleby (ed.), The Domestication and Exploitation of Plants and Animals, London, 1969, 479-484.

FIRTH, R., Malay Fishermen : their Peasant Economy, London, 1946.

--- Primitive Polynesian Economy (2nd ed.), London, 1965.

GAGE, J., 'Gades, l'Inde et les navigations atlantiques dans l'antiquité', Rev.Hist., 205, 1951, 189-216.

GALLANT, T.W., A Fisherman's Tale : an Analysis of the Potential Productivity of Fishing in the Ancient World, (Miscellanea Graeca), Ghent, 1985.

GARCIA DEL TORO, J.R., '"Garum sociorum" : la industria de salazones de pescado en la edad antigua en

Cartagena', Anales de la Universidad de Murcia, 36, 1978, 27-57.

GONZALEZ FERNANDEZ, X., 'O comercio romano na ria de Vigo', Brigantium, 4, 1983, 73-98.

GRIMAL, P. & MONOD, T. 'Sur la véritable nature du "garum"', REA, 54, 1952, 27-38.

HAUSCHILD, T., 'Der römische Leuchtturm von La Coruña (Torre de Hercules). Probleme seiner Rekonstruktion', MDAI(M), 17, 1976, 233-258.

HAUSSOULLIER, B., 'Note sur une inscription de Trézène', Rev. de Phil., n.s., 25, 1901, 336-338.

HESNARD, A., 'Un dépôt augustéen d'amphores à la Longarina, Ostie' in J.H. d'Arms & E.C. Kopff (ed.), The Seaborne Commerce of Ancient Rome (= MAAR, 36), Rome, 1980, 141-156.

HUTTER, S., Der Römische Leuchtturm von La Coruña (Madrider Beitrage, 3), Mainz-am-Rhein, 1973.

IRIA, A., 'Ruinas de tanques de salga de peixe encontradas recentemente em Olhão', Industria Portuguesa, 273, Nov. 1950.

JARDIN, C., 'Garum et sauces de poisson de l'antiquité', RStLig., 27, 1961, 70-96.

JODIN, A., Mogador. Comptoir phénicien du Maroc atlantique (= Etudes et travaux d'archéologie marocaine, 2), Tangiers, 1966.

--- Les établissements du roi Juba II aux Îles Purpuraires (Mogador), Tangiers, 1967.

KAUFMAN WILLIAMS II, C., 'Corinth 1978 : Forum Southwest', Hesperia, 48.2, 1979, 105-44.

KEAY, S., Late Roman Amphorae in the Western Mediterranean : a typology and economic study : the Catalan evidence , 2 vols. (B.A.R. International Series, 196), Oxford, 1984.

KOELHER, C.G., 'Corinthian developments in the study of trade in the fifth century', Hesperia, 50.4, 1981, 449-458.

LAMOUR, C. & MAYET, F., 'Glanes amphoriques. I. Région de Béziers et Narbonnaise', Études sur Pézenas et l'Hérault, 11.1, 1980, 3-16.

LANCHA, J., 'La mosaique d'Océan découverte à Faro (Algarve)', Conimbriga, 24, 1985, 151-175.

LEGRAND, P.E., 'Inscriptions de Trézène', BCH, 24, 1900, 179-215.

LEQUÉMENT, R. & LIOU, B., 'Les épaves de la côte de Transalpine', Cah. Lig. de préh. et d'arch., 24, 1975, 76-82.

LIOU, B. & MARICHAL, R., 'Les inscriptions peintes sur amphores de l'anse Saint-Gervais', Archaeonautica, 2, 1978, 109-181.

LOPEZ F.S., 'La subacuatica en las costas del norte y noroeste peninsular : estado de la cuestion', VI Congreso Internacional de Arqueologia Subacuatica (Cartagena, 1982), Madrid, 1985, 33-58.

MACKIE, I.M., HARDY, R. & HOBBS, G., 'Fermented fish products' in F.A.O. Fisheries Report, No. 100, Rome, 1971, 1-54.

MAIA, M., 'Contribuição para o estudo das anforas romanas de Troia - anforas do tipo "Africano Grande"', Setúbal Arq., 1, 1975, 155-158.

---- 'Cepos de chumbo de ancoras romanas encontradas ao largo de Sesimbra', Setúbal Arq., 1, 1975,

177-180.

---- 'Marcas em anforas da forma Dr. 20 da Tróia, Setúbal', AP, iii, 7-9, 1974-1977, 355-358.

____ 'Anforas neopunicas do sul de Portugal', Actas das III Jornadas Arq., 1977, I, 197-207.

---- 'As anforas de S. Bartolomeu de Castro Marim', Clio [Revista do Centro de Historia da Univ. de Lisboa], 1, 1979, 141-148.

MAGALHÃES, L. de, 'Os barcos da ria de Aveiro', Portugalia, 2.1, 1905-1908, 49-62.

MANACORDA, D., 'Anfore spagnole a Pompei' in L'instrumentum domesticum di Ercolano e Pompei nella prima eta imperiale (= Quaderni di Cultura Materiale, I), Rome, 1977, 121-133.

MARQUES DA COSTA, A.I., 'Estudos sobre algumas estações da epoca luso-romana nos arredores de Setúbal', AP, 27, 1925-26, 165-181.

--- 'Estudos sobre algumas estações da epoca luso-romana nos arredores de Setúbal', AP, 29, 1930-31, 2-31.

MAYET, F., 'Marques d'amphores de Maurétanie Tingitane (Banasa, Thamusida, Volubilis)', MEFRA, 90, 1978, 357-406.

MESQUITA DE FIGUEIREDO, A., 'Contribuição para a historia da pesca em Portugal na epoca luso-romana', AP, 4, 1898, 53-60.

------- 'Ruines d'antiques établissements à salaisons sur le littoral sud du Portugal', Bull.hisp., 8, 1906, 109-121.

MOLINA FAJARDO, F. & LOPEZ CASTRO, J.L., 'Conjunto de monedas de la factoria de salazones de "El Majuelo"', Almuñecar. Arqueologia e Historia, 2, 1984, 257-274.

MORENA PARAMO, A. & ABAD CASAL, L., 'Aportaciones al estudio de la pesca en la antiguedad', Habis, 2, 1971, 209-221.

MROZEK, S., 'Le prix de la pourpre dans l'histoire romaine' in Les dévaluations à Rome. Époque républicaine et impériale II (Gdansk 19-21 oct. 1978) (Collection de l'école française de Rome, 37), Rome, 1980, 235-243.

NAVEIRO LOPEZ, J., 'As anforas romanas a A Coruña (I)', Brigantium, 2, 1981, 117-26.

----- 'As anforas romanas a A Coruña (II)', Brigantium, 3, 1982, 63-74.

PANELLA, C., 'Annotazioni in margine alle stratigrafia della Terme ostiensi del Nuotatore' in Recherches sur les amphores romaines (= Coll. de l'École française de Rome, 10), Rome, 1972, 69-106.

------- 'La distribuzione e i mercati' in A. Giardina & A. Schiavone (ed.), Merci, Mercati e Scambi nel Mediterraneo (= Societa romana e produzione schiavistica, II), Rome/Bari, 1981, 55-80.

PARKER, A.J., 'The evidence provided by underwater archaeology for Roman trade in the western Mediterranean' in D.J. Blackman (ed.), Marine Archaeology (= Colston Papers, 23), Bristol, 1973, 361-381.

----- 'Sicilia e Malta nel commercio marittimo dell' antichita', Kokalos, 22-23, 1976-77, 622-631.

----- 'Lusitanian amphoras' in Méthodes classiques et méthodes formelles dans l'étude des amphores (= Coll. de l'École française de Rome, 32), Paris/Rome, 1977,

35-46.

----- 'Stratification and contamination in ancient Mediterranean shipwrecks', IJNA, 10.4, 1981, 309-335.

---- & PRICE, J., 'Spanish exports of the Claudian period : the significance of the Port Vendres II wreck reconsidered', IJNA, 10.3, 1981, 221-228.

PASCUAL GUASCH, R., 'Algunos aspectos del comercio antiguo segun las anforas' in Comunicaciones a la I Reunión de Historia de la Economia antigua de la Peninsula iberica, Valencia, 1968, 67-79.

--------- 'La evolución de las explotaciones béticas durante el Imperio' in J.M. Blazquez (ed.), Producción y comercio del aceite en la antiguedad, Madrid, 1981, 233-242.

PEACOCK, D.P.S., 'Amphorae and the Baetican fish industry', Antiq. Journ., 54, 1974, 232-243.

-------- 'Roman amphorae : typology, fabric and origin' in Méthodes classiques et méthodes formelles dans l'étude des amphores, Paris/Rome, 1977, 261-278.

PEIXOTO CABRAL, J.M., 'Determinação da proveniencia de ceramicas', Arqueologia (Porto), 4, 1981, 74-82.

PONSICH, M., 'A propos d'une usine de salaison à Belo', MCV, 12, 1976, 69-79.

----- Factores geo-economicos de Bética meridional : el aceite de oliva y las salazones de pescado, Madrid (forthcoming)

----- & Tarradell, M., Garum et industries antiques de salaison dans la Méditerranée occidentale (Bibliothèque de l'École des Hautes Études Hispaniques, XXXVI), Paris, 1965.

PURPURA, G., 'Pesca e stabilimenti antiche per la lavorazione del pesce in Sicilia. I. S. Vito (Trapani), Cala Minnola (Levanzo)', Sicilia Archeologica, 48, 1982, 45-60.

RAMON, J., La producción anforica punico-ebusitana, Madrid, 1981.

RAU, V., A exploração e o comercio do sal de Setúbal (= Estudo de historia economica, 1), Lisbon, 1951.

REESE, D.S., 'The industrial exploitation of Murex shells : purple dye and lime production at Sidi Khrebish, Benghazi (Berenice)', Libyan Studies : 11th Annual Report of the Soc. for Libyan Studies, 1979-80, 79-93.

REMESAL RODRIGUEZ, J., 'Die Ölwirtschaft in der Provinz Baetica : neue Formen der Analyse', Saalburg-Jahrbuch, 38, 1982, 30-71.

RIBEIRO, M., 'Anzóis de Tróia : subsidios para o estudo da pesca no periodo lusitano-romano', AP, iii, 4, 1970, 221-236.

ROCHE, J. & VEIGA FERREIRA, O. da, 'Les fouilles récentes dans les amas coquilliers mésolithiques de Muge (1952-1965)', AP, iii, 1, 1967, 19-41.

RODRIGUEZ ALMEIDA, E., 'Novedades de epigrafia anforaria del Monte Testaccio' in Recherches sur les amphores romaines (= Collection de l'école française de Rome, 10), Rome, 1972, 107-241.

----- Il Monte Testaccio : ambiente, storia, materiali, Rome, 1984.

DOS SANTOS ROCHA, A., 'Estações preromanas da idade do ferro nas vizinhanças da Figueira da Foz em Portugal', Portugalia, 2, 1905-1908, 301-359, 493-516.

SOARES, J., Estação romana de Tróia, Setúbal, 1980.

TCHERNIA, A., 'D. Caecilius Hospitalis et M. Iulius Hermesianus (CIL VI 1625b & 20742)' in J.M. Blazquez (ed.), Producción y comercio del aceite en la antiguedad, Madrid, 1981, 155-160.

------ & Pomey, P., 'Le tonnage maximum des navires de commerce romaines', Archaeonautica, 2, 1978, 233-251.

VALLESPIN, O., 'Carta arqueologica de La Caleta', VI Congreso Internacional de Arqueologia Subacuatica (Cartagena, 1982), Madrid, 1985, 59-74.

------ 'The Copper Wreck (Pecio del Cobre)', IJNA, 15.4, 1986, 305-322.

VASCONCELOS, J. LEITE de, 'Olaria luso-romana em S. Bartolomeu de Castro Marim', AP, 4, 1898, 329-336.

VEIGA FERREIRA, O. da, 'Acerca da presença de "Purpura haemostoma" e "Purpura lapillus" Linné nas estações pré-historicas portuguesas', Rev. Guim., 68, 1958, 377-382.

-------- 'Algumas considerações sobre as fabricas de conservas de peixe na antiguedade encontradas em Portugal', Arq. de Beja, 23-24, 1966-1967, 123-134.

-------- 'Algumas notas acerca da pesca na antiguedade', AP, iii, 2, 1968, 113-133.

VEIGA SILVA PEREIRA, M. L., 'Marcas de oleiros algarvios do periodo romano', AP, iii, 7-9, 1974-1977, 243-268.

VILA VALENTI, J., 'Notas sobre la antigua producción y comercio de la sal en el Mediterraneo occidental', I Congreso arqueologico del Marruecos español, Tetuan 1954, Tetuan, 1954, 225-234.

WEST, L.C., *Roman Imperial Spain, the Objects of Trade*, Oxford, 1929.

WILL, E., 'Le sel des Morins et des Ménapiens' in M. Resnard (ed.), *Hommages à A. Grenier*, III (= Collection Latomus, LVIII), Brussels, 1962, 1649-1657.

WILLIAMS, D.F. & PEACOCK, D.P.S., 'The importation of olive oil into Iron Age and Roman Britain' in *Produccion y comercio del aceite en la antiguedad II*, Madrid, 1983, 263-280.

ZEVI, F., 'Appunti sulle anfore romane', *Arch.Class.*, 18, 1966, 229-247.

INDEX LOCORUM (n.b. Baet. = Baetica; Tarr. = Hispania Tarraconensis)

INDEX NOMINUM

INDEX RERUM